Highway
Materials and
Pavement Testing

Highway Materials and Pavement Testing

DK Maharaj

BSc Engg (Civil), ME, PhD (IIT Madras)
MIGS (Madras Chapter), MIGS (New Delhi, India)

Director, Principal and Professor
Guru Nanak Institute of Technology (GNIT)
Guru Nanak Institutions (GNI)
Mullana, Ambala, Haryana

Former
Director, Principal and Professor AIMT, Dhaurang, Yamuna Nagar, Haryana
GRIMT, Nechron, Radaur, Yamuna Nagar, Haryana
Director, Deputy Director, Dean, Head and Professor
Singhania University, Jhunjhunu, Rajsthan

CBS

CBS Publishers & Distributors Pvt Ltd

New Delhi • Bengaluru • Chennai • Kochi • Kolkata • Mumbai
Bhopal • Bhubaneswar • Hyderabad • Jharkhand • Nagpur • Patna • Pune • Uttarakhand • Dhaka (Bangladesh)

Highway Materials and Pavement Testing

ISBN: 978-93-89185-90-4

Copyright © Author and Publisher

First Edition: 2020

Published by Satish Kumar Jain and produced by Varun Jain for

CBS Publishers & Distributors Pvt Ltd
4819/XI Prahlad Street, 24 Ansari Road, Daryaganj, New Delhi 110 002, India.
Ph: 23289259, 23266861, 23266867   Fax: 011-23243014   Website: www.cbspd.com
e-mail: delhi@cbspd.com; cbspubs@airtelmail.in
Corporate Office: 204 FIE, Industrial Area, Patparganj, Delhi 110 092
Ph: 4934 4934   Fax: 4934 4935   e-mail: publishing@cbspd.com; publicity@cbspd.com

Branches

- **Bengaluru:** Seema House, 2975, 17th Cross, K.R. Road, Banasankari 2nd Stage, Bengaluru 560 070, Karnataka
  Ph: +91-80-26771678/79   Fax: +91-80-26771680   e-mail: bangalore@cbspd.com
- **Chennai:** 7, Subbaraya Street, Shenoy Nagar, Chennai 600 030, Tamil Nadu
  Ph: +91-44-26680620, 26681266   Fax: +91-44-42032115   e-mail: chennai@cbspd.com
- **Kochi:** 42/1325, 1326, Power House Road, Opposite KSEB Power House, Ernakulam 682 018, Kochi, Kerala
  Ph: +91-484-4059061-65   Fax: +91-484-4059065   e-mail: kochi@cbspd.com
- **Kolkata:** 6/B, Ground Floor, Rameswar Shaw Road, Kolkata-700 014, West Bengal
  Ph: +91-33-22891126, 22891127, 22891128   e-mail: kolkata@cbspd.com
- **Mumbai:** 83-C, Dr E Moses Road, Worli, Mumbai-400018, Maharashtra
  Ph: +91-22-24902340/41   Fax: +91-22-24902342   e-mail: mumbai@cbspd.com

Representatives

• Bhopal	0-8319310552	• Bhubaneswar	0-9911037372	• Hyderabad	0-9885175004
• Jharkhand	0-9811541605	• Nagpur	0-9421945513	• Patna	0-9334159340
• Pune	0-9623451994	• Uttarakhand	0-9716462459	• Dhaka (Bangladesh)	01912-003485

Printed at: Mudrak, Noida, UP, India

to
my parents

Preface

The book *Highway Materials and Pavement Testing* has been written for undergraduate and postgraduate students of civil engineering and engineers to cover the various highway material testing and field testing in highway engineering. Each chapter of the book is clear and sufficient in itself.

Feedback and suggestions from the readers of this book are welcome in improving the future printings | editions of the book.

DK Maharaj

Acknowledgements

I express my deep sense of gratitude to my parents for inspiration which I got from them. I express my sincere thanks to Chairman, Mr Tralochan Singh, and Dean, Dr Sachin Chawla, for their suggestions. I sincerely thank the authorities of various universities and institutions for giving me this course to teach. I sincerely thank the team members of CBS Publishers & Distributors for bringing out this book in present form. Last but not the least, I wish to express my gratitude to my wife and loving sons Ashish and Manish for extending all the cooperation during the preparation of this book.

DK Maharaj

Contents

PART 1

Tests on Soil Subgrade

CHAPTER 1

Grain Size Analysis

SIEVE ANALYSIS

1.1 OBJECT

To determine grain size distribution of a soil by sieve analysis.

1.2 THEORY

The grain size analysis is also known as mechanical analysis. In this analysis the percentage of individual grain sizes present is determined by sieving a known weight of soil through successive smaller sieves. Based on grain size the soil is divided into four parts as follows:

Gravel: Fraction of soil >4.75 mm

Sand: Size 0.075–4.75 mm

Silt: Size 0.002–0.075 mm

Clay: <0.002 mm

1.3 APPLICATIONS

The results of grain size distribution are widely used for soil classification, design of filters, construction of earth dams, highway embankments, for construction of building, hydraulic structures and road construction, etc.

1.4 APPARATUS

- For coarse sieve analysis: IS sieves 100, 63, 20, 10 and 4.75 mm.
- For fine sieve analysis: IS sieves 2, 1, 0.6, 0.425, 0.212, 0.150 and 0.75 mm.
- Oven
- Balance accurate to 0.1 g
- Weights
- Sieve shaker (Fig. 1.1)
- Tray
- Pan

Fig. 1.1: Sieve shaker (grain size analysis)
(*Courtesy:* AIMIL Ltd.)

1.5 COARSE SIEVE ANALYSIS

- Take suitable quantity of oven-dry soil.
- Arrange the first set of sieves such that 100 mm sieve is at the top and 4.75 mm sieve is at the bottom.
- Put cover on top sieve and pan at the bottom of 4.75 mm sieve and put the soil on top sieve before covering it.
- Put the sieves in the sieve shaker and clamp it tightly.
- Shake the sieves for 10 minutes.
- Find the weight of soil retained on each sieve.
- Find the weight of soil on pan.

1.6 FINE SIEVE ANALYSIS

- Arrange the second set of sieves such that 2 mm sieve is at the top and 0.075 mm sieve is at the bottom.
- Put pan at the bottom of 0.075 mm sieve.
- Put the soil passing 4.75 mm sieve on the top sieve and then cover it.
- Put the set of sieves with pan and cover in the sieve shaker.
- Shake the sieves for 10 minutes.
- Find the weight of soil retained on each sieve.

1.7 WET SIEVE ANALYSIS

Wet sieve analysis is considered for clayey or cohesive soil:
- Take soil finer than 2 mm size and oven-dry it 105–110 C.
- Spread the sample in a tray and cover it with water.
- Stir the mix and leave for soaking.
- Put the soaked soil specimen on the top sieve of the set of sieves such that the finest sieve and pan is at the bottom.
- Wash the soaked soil specimen thoroughly. Continue washing till the water passing each sieve is substantially clean.
- Empty the fraction of soil on each sieve carefully. Dry the soil in oven at 105–110 C.
- Weigh the oven dry-soil separately.

Note: The sieve sizes for wet sieve analysis are same as the fine sieve analysis.

For coarse grained analysis, fine grained analysis and wet sieve analysis, find the percentage of soil retained on each sieve; cumulative percentage retained and percentage finer. Particle size distribution curve is obtained by plotting particle size on x-axis on log scale and percentage finer on *y*-axis.

1.8 PRECAUTIONS

- The temperature of oven should be between 105–110 C.
- The soil should not come out while shaking the sieves in the sieve shaker.
- Determine percentage finer with respect to total soil taken.

1.9 OBSERVATIONS AND CALCULATIONS

Table 1.1 shows coarse sieve analysis.

Table 1.1: Coarse sieve analysis

Weight of soil taken for analysis = ... g

1st sieve size (mm)	Particle size (mm)	Weight of soil retained (g)	Percentage weight retained	Cumulatf-e retained percentage	Percentage finer (NJ)
100					
63					
40					
20					
4.75					

Table 1.2 shows fine sieve analysis and wet sieve analysis.

Table 1.2: Fine sieve analysis and wet sieve analysis

Weight of soil taken for analysis = ... g

Ist Sieve	Particle Size (mm)	Wight of soil retained (g)	Percentage weight retained	Cumulative Percentage retained	Percentage finer (N)
2					
1					
0.600					
0.425					
0.212					
0.150					
0.0075					

1.10 RESULT

1. Plot curve between percentage finer and grain size on semi log graph
2. Find particle size for 10% finer, D_{10}: Particle size for 30% finer, D_{30}: Particle size for 60% finer, D_{60}.

3. Find uniformity coefficient $Cu = \dfrac{D_{60}}{D_{10}}$

4. Find coefficient of curvature $Cc = \dfrac{(D_{30})^2}{(D_{60} \times D_{10})}$

1.11 GRAIN SIZE ANALYSIS FOR SOIL PASSING 0.075 MM SIEVE BY SEDIMENTATION

The following are the methods used to find the grain size distribution of soils smaller than 0.075 mm sieve.

1. Pipet method
2. Hydrometer method

Details of these methods are not included as it is beyond the scope of coverage. For details please refer author's "Laboratory Manual for Soil Testing" and IS 2720 part IV, Indian Standard Methods of Test for Soils, Grain Size Analysis.

1. What is the purpose of sieve analysis?
2. Define coefficient of curvature (Cc) and coefficient of uniformity (Cu). What are their applications?
3. What do you understand by uniformly graded and well graded soil?
4. Write the meaning of soils which are GW, GM, GC, SW, SP, SM, SC, SW-SM, GP-GC.
5. Why do you use semilog graph paper for plotting grain-size distribution curve?
6. What are the precautions in sieve analysis?
7. What are the sources of errors in sieve analysis?
8. How do you decide the type of samples to be taken for sieve analysis?
9. What are the poorly graded, well graded and uniformly graded soil?
10. What is the difference between dry sieve analysis and wet sieve analysis?

CHAPTER 2

Liquid Limit and Plastic Limit

2.1 THEORY

The limiting water content when a soil mass passes from liquid to plastic state of consistency is termed liquid limit. Liquid limit is the water content at which a part of soil, cut by a groove of standard dimensions, will flow together for a distance of 13 mm under an impact of 25 blows in Casagrande's liquid limit apparatus. The limiting water content when a soil mass passes from plastic to semisolid state of consistency is termed plastic limit. Plastic limit is defined as the water content at which a soil will just begin to crumble when rolled into a thread of 3 mm in diameter.

2.2 APPLICATIONS

Fine grained cohesive soils are classified by knowing liquid limit and plastic limit. From liquid limit and plastic limit we can find flow index, toughness index and plasticity index. These give an idea about plasticity, cohesiveness, compressibility, shear strength and permeability of cohesive soils.

2.3 LIQUID LIMIT TEST

2.3.1 Object

To determine liquid limit of soil.

2.3.2 Apparatus

- Casagrande's liquid limit apparatus (Fig. 2.1)
- Grooving tool
- Spatula
- Mixing dish or bowl
- Containers for water content
- Balance, sensitive to 0.01 g
- Oven
- Sieve 0.425 mm.

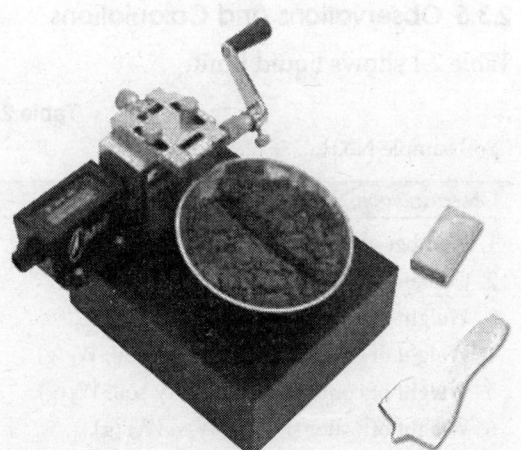

Fig. 2.1: Liquid limit test apparatus
(*Courtesy:* AIMIL Ltd.)

2.3.3 Procedure

- Clean, dry and make oil free the bowl of liquid limit device.
- Keep the height of drop of bowl equal to 10 mm.

- Take about 250 g of dry soil sample passing 0.425 mm sieve.
- Mix the soil with water on a glass plate with a spatula until the mixture is uniform and behaves as a soft paste.
- Place 50 to 80 g of soil paste in the bowl and level it off to a depth of approximately 1 cm, i.e. 10 mm.
- Cut a groove through the sample from back to front dividing the paste in the bowl into two equal halves. Consider Casagrande's tool for a normal fine grained soil and ASTM tool for sandy fine grained soil.
- Turn the handle of Casagrande's device at a steady rate of two revolutions per second. Continue turning until two halves of soil pat come in contact at the bottom of the groove along a distance of 13 mm. Note the number of blows. The groove should come in contact due to flow of soil not due to sliding.
- Take 5 to 10 g of soil from the sample and put it in a small container for water content determination.
- Repeat steps 3 to 8 for four to five times. Note down the number of drops and water content each time. It is recommended that the water content should be varied such that the number of drops are between 15 and 30.

2.3.4 Precautions

- The soil used in liquid limit test should not be oven dried.
- The groove made in the soil in liquid limit test should close 13 mm by flow of soil from either side and not by slippage.
- The lid of container should be made tight immediately after putting the soil in the container.
- Add water in the different soil sample such that the number of blows (drops) range from 15 to 35

2.3.5 Observations and Calculations

Table 2.1 shows liquid limit.

Table 2.1: Liquid limit

Soil sample No.:........................   Date : ....

Observation No.	1	2	3
1. Number of drops, N			
2. Container number			
3. Weight of container with lid, W_1 (g)			
4. Weight of container + lid + wet soil, W_2 (g)			
5. Weight of container + lid + dry soil, W_3 (g)			
6. Weight of water $(W_w) = (W_2 - W_3)$ (g)			
7. Weight of dry soil $(W_d) = (W_3 - W_1)$ (g)			
8. Water content $w = \dfrac{W_2 - W_3}{W_3 - W_1} \times 100\%$ $= \dfrac{W_w}{W_d} \times 100\%$			

Plot log N on x-axis and water content, w on y-axis. From plot find the water content for 25 numbers of blows (drops). This gives liquid limit (Fig. 2.2).

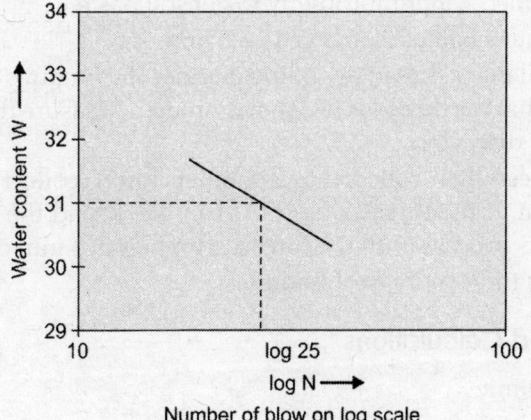

Fig. 2.2: Liquid limit result

1. Define liquid limit. What do you understand by it?
2. To determine liquid limit in the laboratory, how is it defined?
3. At liquid limit, what will be the consistency of soil?
4. At liquid limit, what will be the state of soil?
5. At liquid limit, what will be the shear strength of soil?
6. Explain giving reasons as to why liquid limits of 2 samples from two sites are different?
7. Describe in brief the applications of liquid limit.
8. What is the name of liquid limit apparatus? Who invented it?
9. Whether undisturbed or remoulded sample is used to determine liquid limit and why?
10. In order to estimate liquid limit plane graph paper or semilog graph paper is used and why?
11. Define flow curve and flow index.

2.4 PLASTIC LIMIT TEST

2.4.1 Object

To determine plastic limit of soil.

2.4.2 Apparatus

• Glass or plastic plate
• Metal rod of 3 mm diameter
• Spatula
• Containers (small size)
• Balance sensitive to 0.01 g
• Oven.

2.4.3 Procedure

- Take about 20 g of soil for plastic limit test. The soil should pass through 0.425 mm IS sieve.
- Mix the soil with distilled water thoroughly to get soil paste.
- Make the soil paste into a ball of diameter 10–20 mm.
- Convert the ball of soil into a thread by rolling it under the fingures against the glass surface. Roll the thread such that the thread is of 3 mm diameter. Measure the diameter of the thread by metal rod of 3 mm diameter.
- If the thread crumbles when rolled into diameter 3 mm, collect such threads for water content determination. If the thread does not crumble, knead the sample and again make the thread. Repeat this process until the thread crumbles at 3 mm diameter.
- Repeat steps 1–5 with three more fresh samples.

2.4.4 Observations and Calculations

Table 2.2 shows plastic limit.

Table 2.2: Plastic limit

Soil sample No.:............................   Date : ....

Observation No.	1	2	3
1. Container number			
2. Weight of container with lid, W_1 (g)			
3. Weight of container + lid + wet soil, W_2 (g)			
4. Weight of container + lid + dry soil, W_3 (g)			
5. Weight of dry soil $(W_d) = (W_3 - W_1)$ (g)			
6. Weight of water $(W_w) = (W_2 - W_3)$ (g)			
7. Plastic limit $= \dfrac{W_w}{W_d} \times 100\%$ (water content)			

$$\text{Average plastic limit,} \quad W_p = \dots\dots \, (\%)$$

$$\text{Natural water content,} \quad W = \dots\dots \, (\%)$$

$$\text{Plasticity index } (I_p) = \text{Liquid limit} - \text{Plastic limit}$$

$$= (W_L - W_p) \dots\% \tag{2.1}$$

$$\text{Consistency index} = \frac{W_L - w}{I_P} \tag{2.2}$$

$$\text{Liquidity index} = \frac{w - W_p}{I_P} \tag{2.3}$$

QUESTIONS

1. Define plastic limit. What do you understand by it?
2. To determine plastic limit in the laboratory how is it defined?
3. At plastic limit, what will be the consistency of soil?

4. At plastic limit, what will be the state of soil?
5. Do different soils have different consistency at their plastic limits?
6. What would be the diameter of soil while determining plastic limit and why?
7. How will the plastic limit be affected if the diameter of thread is >3 mm?
8. Will the two soil samples have the same shear strength at their plastic limits? Explain it.
9. Whether the plastic limit will be same or different for two soils obtained from the same site.

CHAPTER 3

Shrinkage Factors of Soil

3.1 OBJECT

To determine the shrinkage factors, i.e. shrinkage limit, shrinkage ratio, shrinkage index and volumetric shrinkage of soils.

3.2 THEORY

The shrinkage limit is the maximum water content at which a further reduction in water content will not cause a decrease in volume of the soil mass. The shrinkage limit is obtained as

$$W_s = W_s = \left[w - \frac{(V_w - V_d)}{W_d} \times 100 \right] \%$$ (3.1)

Where w is the water content

V_w is the internal volume of shrinkage dish

v_d is the volume of dry soil pat

W_d is the weight of dried soil pat

Shrinkage index, $I_s = (I_p - W_s)$ (3.2)

Shrinkage ration, $R = \dfrac{W_d}{V_d}$ (3.3)

Volumetric shrinkage, $V_s = (W_1 - W_s)R$ (3.4)

$$= (W_1 - W_s)\frac{W_d}{V_d}$$ (3.5)

Where W_1 is the given water content in percentage.

3.3 APPLICATIONS

By knowing the shrinkage limit we understand the swelling and shrinkage properties of cohesive soils. From shrinkage limit we find the suitability of soil as a construction material in foundations, roads, embankments and dams.

3.4 APPARATUS

Shrinkage dish 45 mm diameter and 15 mm height (Fig. 3.1).
- Evaporating dish
- Glass cup 50 to 55 mm in diameter and 25 mm in height
- Glass plates, plain and with metal prongs 75 75 mm, 3 mm thick
- Spatula
- Straight edge 150 mm length
- Balance of 0.1 g sensitivity
- Mercury
- Drying oven.

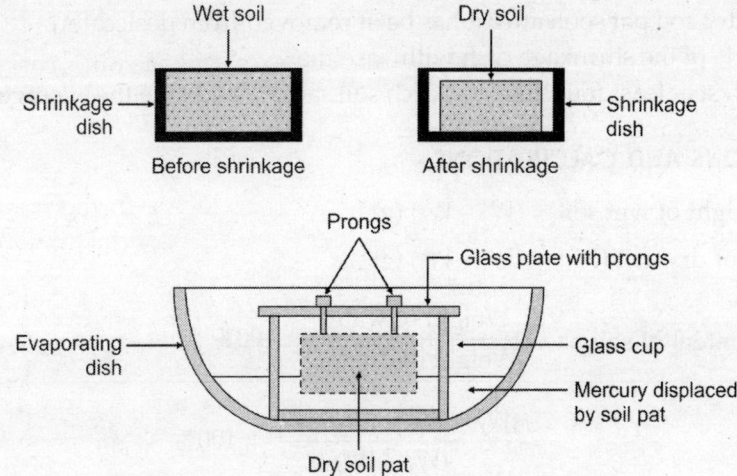

Fig. 3.1: Shows the apparatus for shrinkage limit

3.5 PROCEDURE

1. Weight the shrinkage dish. Let it be W_1. Put the shrinkage dish in an evaporating dish. Fill the shrinkage dish with mercury. Put a glass plate on the top of the shrinkage dish and press it. Find the weight of shrinkage dish with mercury. Then find the weight of mercury in shrinkage dish. Divide the weight of mercury with its unit weight. It gives the internal volume of shrinkage dish. Let it be V_w.

2. Mix 50 gram of soil passing 425 micron, i.e. 0.425 mm sieve with distilled water in an evaporating dish. The water added should be somewhat greater than the liquid limit.

3. Coat the inside of shrinkage dish with a thin layer of grease.

4. Fill the dish by soil paste up to one-third of its height. Tap the dish so that soil paste flow towards end. Repeat this for two more layers and fill the shrinkage dish. Strike off the excess soil paste with straight edge; clean outside surface and find the weight of shrinkage dish with wet soil. Let it be W_2.

5. Dry the soil in air until the colour of soil pat turns light. Then dry it in a temperature controlled oven at 105–110 C. After drying, cool it in air and then find the weight of shrinkage dish with dry soil pat. Let it be W_3.

6. Fill the glass cup with mercury.
7. Place the glass cup with mercury in a large evaporating dish.
8. Place dry soil pat on the surface of mercury.
9. With the help of glass plate with prongs, press the soil pat such that it is completely submerged in mercury.
10. Find the weight of displaced mercury. Divide it by unit weight of mercury. This gives the volume of dry soil pat. Let it be V_d.

3.6 PRECAUTIONS

1. While filling the shrinkage dish with soil paste, do sufficient taping to remove the entrapped air.
2. Weigh the dry soil pat soon after it has been removed from desiccator.
3. Grease inside of the shrinkage dish with vaseline.
4. Repeat the test at least four times for each soil sample and take the average.

3.7 OBSERVATIONS AND CALCULATIONS

Weight of wet soil $= (W_2 - W_1)$ (g)

Weight of dry soil $(W_d) = (W_3 - W_1)$ (g)

Water content of soil, $w = \dfrac{\text{Weight of water}}{\text{Weight of dry soil}}$ 100%

$$= \frac{(W_2 - W_1) - (W_3 - W_1)}{(W_3 - W_1)} \times 100\% \qquad (3.6)$$

Shrinkage limit (remoulded soil)

$$W_s = \left[w - \frac{(V_w - V_d)}{W_d} \times 100 \right]\%$$

Shrinkage index, $\quad Is = Ip - Ws$

Shrinkage ratio, $\quad R = \dfrac{W_d}{V_d}$

Volumetric shrinkage, $V_s = (W_8 - W_s)\, R$

$$= (W_8 - W_s)\, \frac{W_d}{V_d}$$

Where W_8 is the given water content in percentage.

Table 3.1 shows shrinkage limit.

Table 3.1: Shrinkage limit

Soil sample No.:............................   Date : ....

Observation No.	1	2	3	4
1. Weight of shrinkage dish, W_1 (g)				
2. Weight of shrinkage dish with wet soil pat, W_2 (g)				
3. Weight of shrinkage dish with dry soil pat, W_3 (g)				
4. Weight of dry soil pat, $W_d = (W_3 - W_1)$ (g)				
5. Weight of wet soil pat, $(W_{wet}) = (W_2 - W_1)$ (g)				
6. Volume of shrinkage dish = Volume of wet soil, V_w (cc)				
7. Volume of dry soil = V_d (cc)				
8. Water content, $w = \dfrac{[(W_2 - W_1) - (W_3 - W_1)]}{W_3 - W_1} \times 100\%$				
9. Shinkage Limit, $= \left[w - \dfrac{(V_w - V_d)}{W_d} \times 100 \right] \%$				

QUESTIONS

1. What do you understand by shrinkage limit?
2. If water content is reduced at shrinkage limit what is the reason for no decrease of volume of soil?
3. Write the factors which affect the value of shrinkage limit.
4. Write the practical application of shrinkage factors.
5. At shrinkage limit, what is the consistency of soil?
6. At shrinkage limit, what is the degree of saturation?
7. If water is added to the soil sample at shrinkage limit, what will be its effect on volume of soil?
8. What is the importance of coating inside the shrinkage dish with grease?
9. Three prongs are provided on the glass plate, what is the reason for this?
10. Whether undisturbed or remoulded sample is used in the shrinkage limit test?

CHAPTER 4

In Situ Density (Core Cutter Method and Sand Replacement Method)

4.1 THEORY

Density is defined as mass per unit volume of soil. The density can be expressed in g/cm^3, kg/m^3, etc.

Density of Soil

$$r = \frac{W}{V}$$

where r is density of soil, V is the volume of soil, W is mass of soil.

Wet Density of Soil

Wet density of soil is the mass of wet soil per unit volume of soil, i.e.

$$r_b = \frac{W_{wet}}{V}$$

Where r_b is the wet density (bulk density) of soil and V is the volume of soil.

Dry Density of Soil

Dry density of soil, $r_d = \dfrac{r_b}{1+w}$

where w is water content.

Void Ratio

Void ratio is the ratio of volume of voids to volume of soil solids.

$$\text{Void ratio, } e = \frac{V_v}{V_s} \times 100$$

where V_v is the volume of void and V_s is the volume of soil solids.

Degree of Saturation

Degree of saturation is defined as the ratio of volume of water to volume of voids.

i.e., Degree of saturation, $S = \dfrac{V_W}{V_V} \times 100$

where V_w = Volume of water and V_v is the volume of void.

4.2 APPLICATIONS

With the help of density of soil we can find bearing capacity of soil foundation system, settlement of footings, earth pressure behind retaining walls, dams and embankments. The density also helps in checking the stability of dams, natural slopes and embankments. It helps in determining the relative density of soil.

4.3 CORE CUTTER METHOD

4.3.1 Object

To determine the field density by core cutter method.

4.3.2 Apparatus

- Cylindrical core cutter of diameter 10 cm and height 12.74 cm
- Steel dolly
- Balance
- Straight edge
- Knife
- Water content cans
- Oven
- Steel rammer.

4.3.3 Procedure

1. Find the weight of the core cutter; Let it be W_1.
2. Measure internal diameter and height of the core cutter. Find internal volume of core cutter. Let it be V_0.
3. In the field, clean and level the ground where density is to be determined.
4. Push the cylindrical core cutter into the soil to its complete depth by gently ramming it by rammer.
5. Excavate the soil around the core cutter and remove the excavated soil.
6. Lift the core cutter up carefully so that no soil comes out of the core cutter. The soil must be projected up and down of the core cutter.
7. Trim the bottom and top surface of the sample very carefully.
8. After cleaning the outside of the core cutter, determine the weight of core cutter with soil. Let it be W_2.
9. Take the empty weight of can. Let it be W_3.
10. After removing the soil from core cutter put some soil in the can for water content determination. Let the weight of can with soil be W_4.
11. Put the can with wet soil in oven at 105–110 C for 24 hours for drying.
12. Find the weight of dry soil with can. Let it be W_5.

4.3.4 Precautions

1. The core cutter should be removed gently by putting steel dooly on the core cutter.
2. Excavate the soil all around the core cutter before lifting the cutter.
3. Take care that no soil drops down while taking out the cutter from soil.

4. The soil for water content determination should be kept in oven at temperature 105–110 C.

4.3.5 Observations and Calculations

Weight of core cutter, $W_1 = ...(g)$

Weight of core cutter with soil, $W_2 = ...(g)$

Weight of wet soil, $W_{wet} = W_2 - W_1 = ...(g)$

Volume of soil = Internal volume of core cutter

$$= V_0 = ...(cm^3)$$

Bulk density of soil $(r_b) = \dfrac{W_{wet}}{V_0}$

$$= \dfrac{W_2 - W_1}{V_0} ....(g/cm^3)$$

Weight of can = $W_3 = ...(g)$

Weight of can + wet soil = $W_4 = ...(g)$

Weight of can + dry soil = $W_5 = ...(g)$

Weight of water $(W_w) = W_4 - W_5 = ...(g)$

Weight of dry soil = $W_d = W_5 - W_3 = ...(g)$

Water content, $w = \dfrac{\text{Weight of water}}{\text{Weight of dry soil}} \times 100$

$$= \dfrac{W_w}{W_d} \times 100\% \qquad (4.1)$$

Dry density of soil, $r_d = \dfrac{r_b}{1 + w} = ...(g/cm^3)$ $\qquad (4.2)$

Repeat the above steps for 3 to 4 samples for the same levelled ground.

4.4 SAND REPLACEMENT METHOD

4.4.1 Object

To determine field density of soil by sand replacement method.

4.4.2 Apparatus

- Sand pouring cylinder
- 30 cm^2 metal tray with 10 cm hole in the centre.
- Cylindrical calibration container
- Balance
- Cans

- Oven
- Glass plate
- Clean oven dried sand passing 600 micron sieve, i.e. 0.600 mm sieve.

Figure 4.1 shows the sand pouring cylinder, calibrating container and metal tray with hole in the centre.

Fig. 4.1: *In situ* density by sand replacement method
(*Courtesy:* AIMIL Ltd.)

4.4.3 Produce

I. Determination of Density of Sand in Laboratory

1. Measure the internal diameter and height of the calibrating container and find its volume. Let it be V_c.
2. Find the weight of sand pouring cylinder filled with sand. Let it be W_1.
3. Place the sand pouring cylinder on glass plate and open its shutter. The sand falls and fills the cone of sand pouring cylinder. Close the shutter when cone is completely filled by sand. Find the weight of sand pouring cylinder with remaining sand. Let it be W_2.
4. Put the sand pouring cylinder concentrically on the top of the calibrating container. Open the shutter. Sand falls and fills the calibrating container and cone completely. Close the shutter. Find the weight of sand pouring cylinder with remaining sand. Let it be W_3.
5. Find the density of sand.

II. Determination of Water Content

1. Weigh the metal tray having central hole. Let the weight be W_1.
2. Place the metal tray with central hole on levelled ground.
3. Excavate soil of diameter equal to the diameter of hole and depth approximately 15 cm.
4. Put the excavated soil in tray and find its weight. Let the weight of tray and soil be W_2.
5. Put the tray with soil in oven. Find the weight of tray with dry soil. Let it be W_3.
6. Determine water content.

III. Density of Soil in the Field

1. Fill the sand pouring cylinder by sand and weigh it. Let the weight be W_1.
2. Put the pouring cylinder over the hole, and open the shutter until the sand fills completely the hole and cone. Close the shutter. Find the weight of sand pouring cylinder with remaining sand. Let it be W_2.
3. Subtract the weight of sand in cone to $W_1 - W_2$. It gives the weight of sand filled in the hole. Let it be W_{hole}.

4. Find the volume of hole by dividing the weight of sand filled in the hole by density of sand. Let this volume be V_{hole}.

5. Find density of soil, dividing the weight of soil excavated from hole by volume of hole. It gives the density of soil in field. Let the density by r_b.

4.4.4 Precautions

1. Do not leave loose material in the hole.
2. There should be no vibration.
3. Take average value of density as density varies from point to point.
4. In no case the side of hole should cave in.

4.4.5 Observations and Calculations

Table 4.1 shows determination of density of sand in laboratory.

Table 4.1: Determination of density of sand in laboratory

Soil sample No.:............................   Date : ....

Observation No.	1	2	3
1. Volume of calibrating container, V_c (cm^3)			
2. Weight of pouring cylinder + sand, W_1 (g)			
3. Weight of pouring cylinder with remaining sand, W_2(g) (after filling the cone)			
4. Weight of pouring cylinder with remaining sand, W_3(g) (after filling the calibrating container and cone)			
5. Weight of sand filled in cone = $(W_1 - W_2)$ (g)			
6. Weight of sand filled in cylinder and cone = $(W_2 - W_3)$ (g)			
7. Weight of sand in calibrating container $Wc = [(W_2 - W_3) - (W_1 - W_2)]$(g)			
8. Hence, density of sand $r_d = \dfrac{W_c}{V_c}$ (g/cm$_3$)			

Table 4.2 shows water content determination.

Table 4.2: Water content determination

Soil sample No.:............................   Date : ....

Observation No.	1	2	3	4
1. Tray No.				
2. Weight of tray = W'_1 (g)				
3. Weight of tray with excavated wet soil = W'_2(g)				
4. Weight of tray with dry soil = W'_3(g)				
5. Weight of water $(W_w) = (W'_2 - W'_3)$ (g)				
6. Weight of dry soil $(W_d) = (W'_3 - W'_1)$ (g)				
7. Water content, $w = \dfrac{W_w}{W_d} \times 100\%$				

Table 4.3 shows field density of soil.

Table 4.3: Field density of soil

Soil sample No.:........................... Date : ....

Observation No.	1	2	3	4
1. Weight of pouring cylinder filled with sand = W_1 (g)				
2. Weight of pouring cylinder with sand after filling the hole and cone = W_2(g)				
3. Weight of sand filled in cone = W_{cone} (Refer Table 4.1)				
4. Weight of sand filled in hole (W_{hole}) = $[(W_1 - W_2) - W_{cone}]$ (g)				
5. Volume of hole V_{hole} (cm³) = $\dfrac{W_{hole}}{r_d}$ where r_d is density of sand (refer Table 4.1)				
6. Hence, density of soil, r_b (g/cm³) = $\dfrac{\text{Weight of excavated soil}}{\text{Volume of hole}} = \dfrac{W_2 - W_1}{V_{hole}}$ (Refer Table 4.2)				

QUESTIONS

1. What are dry density, wet density and saturated densities of a soil?
2. Define submerged density.
3. Out of dry density, wet density and saturated densities, which density is maximum and which one is minimum?
4. Compare density, relative density and specific gravity of a soil.
5. Describe in brief the precautions to be taken in core cutter method.
6. Describe in brief the precautions to be taken in sand replacement method.
7. Differentiate between core cutter method and sand replacement method.
8. How the density of soil gets affected when the core cutter is inserted in the soil?

CHAPTER 5

Compaction Test

5.1 OBJECT

To determine the optimum moisture content and maximum dry density of a soil by standard and modified proctor tests.

5.2 THEORY

In compaction the soil gets densified due to the reduction of air voids. The degree of compaction of a soil is measured in terms of its dry density. The degree of compaction mainly depends upon its moisture content, compaction energy and type of soil. Compaction energy remaining the same each soil attains the maximum dry density at a particular water content. This water content is known as optimum water content.

5.3 APPLICATIONS

Due to compaction, the density, shear strength and bearing capacity of soil increase. The result of compaction is to reduce void ratio, porosity, permeability and settlements. The stability of earthen dams, embankments, roads, etc. are achieved from results of compaction tests. The laboratory results are useful for field problems.

5.4 APPARATUS

Cylindrical mould of capacity 1000 ml having internal diameter 100 mm and height 127.3 mm or cylindrical mould of capacity 2250 ml having internal diameter 150 mm and height 127.3 mm (Fig. 5.1).

The cylindrical mould is having base plate and removable extension collar:
- Rammer for light compaction, weight 2.6 kg and free drop 310 mm.
- Rammer for heavy compaction, weight 4.89 kg and free drop 450 mm.
- IS sieves: 20 mm and 4.75 mm.
- Oven to maintain a temperature of 105 to 110°C.
- Balance 10 kg capacity with 1 g accuracy.
- Balance 200 g capacity with accuracy 0.01 g.
- Spatula.
- Steel straight edge.

Fig. 5.1: Shows the apparatus for compaction test
(Courtesy: AIMIL Ltd.)

5.5 PROCEDURE

1. Take about 20 kg air dried soil sample for 1000 ml mould or about 40 kg for 2250 ml mould. Sieve the soil through 20 mm and 4.75 mm IS sieve. If the percentage of soil retained on 4.75 mm sieve is less than 20, take 1000 ml mould and if percentage retained on 4.75 mm IS sieve is greater than 20, take 2250 ml mould.

2. Mix the soil passing 4.75 mm IS sieve and retained on 4.75 mm IS sieve.

3. Take about 2.5 kg of soil for 1000 ml mould and 5.6 kg for 2250 ml mould for light compaction. Similarly take 2.8 kg of soil for 1000 ml mould and 6 kg for 2250 ml mould for heavy compaction.

4. Add water about 4% for coarse grained soil and 8% for fine grained soil. Keep this soil water mix in an airtight container for about 5 to 30 minutes for sandy soil and 18–20 hours for clayey soils.

5. Find the weight of mould with base plate. Let it be W_1.

6. Fix the collar on the mould and apply grease inside of the mould and collar.

7. For light compaction, use 2.6 kg rammer with height of fall 310 mm with 25 number of blows for a layer of soil. Scratch the compacted soil on top and put the second layer and compact it as the first layer. Similarly scratch the top of second layer and put the soil on it and compact it as the first layer. In case of 2250 ml mould compaction will be similar to mould of 1000 ml except that the number of blows now will be 56.

8. For heavy compaction, the compaction is done using 4.89 kg hammer with free fall of 450 mm in five layers. Each layer is compacted by 25 blows for 1000 ml mould and 56 blows for 2250 ml mould.

9. Remove the collar and carefully level off the top of the mould by means of straight edge. find the weight of mould plus base plate plus wet soil. Let it be W_2.

10. Eject the soil from the mould and cut at the middle and take samples for water content determination.

11. Repeat steps 6–10 for 4–5 times, using a fresh part of soil specimen each time. Add water to the specimen such that the water content increases each time.

5.6 PRECAUTIONS

1. During compaction, the mould should be placed on a solid base.
2. The compaction blows should be uniformly distributed over the surface of each layer.
3. Scratch each layer of compacted soil and then put the next layer and compact.
4. After compacting the last layer the soil should project about 5 mm above the top rim of the mould.

5.7 OBSERVATIONS AND CALCULATIONS

Table 5.1 shows standard proctor test.

Table 5.1: Standard proctor test

Soil sample No.:.......................... Date : ....
Volume of mould (V) = 945 cc Weight of rammer = 2.5 kg
Number of layers = 3 Number of blows = 25

Observation No.	1	2	3	4	5
Determination of bulk density of soil, r_b					
1. Weight of mould + base plate, W_1 (g)					
2. Weight of mould + base plate + compacted soil, W_2 (g)					
3. Weight of compacted soil (w), = $(W_2 - W_1)$ (g)					
4. Bulk density $r_b = \dfrac{W}{V} \, \text{g/cm}^3$					

Table 5.2 shows standard proctor test determination of dry density.

Table 5.2: Standard proctor test (determination of dry density)

Observation No.	1	2	3	4	5
Determination of water content					
1. Can no.					
2. Weight of can + lid, W_1 (g)					
3. Weight of can + lid + wet soil, W_2 (g)					
4. Weight of can + lid + dry soil, W_3 (g)					
5. Weight of water = $(W_2 - W_3)$ (g)					
6. Weight of dry soil $W_d = (W_3 - W_1)$ (g)					
7. Water content $w = \dfrac{W_w}{W_d} \times 100\%$					
8. Dry density $= \dfrac{r_b}{1+w} \, \text{g/cm}^3$					

Table 5.3 shows modified proctor compaction

Table 5.3: Modified proctor compaction

Soil sample No.:..............................   Date : ....
Volume of mould (V) = 945 cc   Weight of rammer = 4.5 kg
Number of layers = 5   Number of blows = 25

Observation No.	1	2	3	4	5
Determination of bulk density of soil, r_b					
1. Weight of mould + base plate, W_1 (g)					
2. Weight of mould + base plate + compacted soil, W_2 (g)					
3. Weight of compacted soil (w)=($W_2 - W_1$) (g)					
4. Bulk Density $r_b = \dfrac{W}{V}$ g/cm^3					

Table 5.4 shows modified proctor test (determination of dry density).

Table 5.4: Modified proctor test (determination of dry density)

Observation No.	1	2	3	4	5
Determination of water content					
1. Can no.					
2. Weight of can + lid, W_1 (g)					
3. Weight of can + lid + wet soil, W_2 (g)					
4. Weight of can + lid + dry soil, W_3 (g)					
5. Weight of water = ($W_2 - W_3$) (g)					
6. Weight of dry soil $Wd = (W_3 - W_1)$ (g)					
7. Water content $w = \dfrac{W_w}{W_d} \times 100\%$					
8. Dry density = $\dfrac{r_b}{1+w}$ g/cm^3					

QUESTIONS

1. What do you mean by compaction of soil?
2. What do you mean by optimum water content? Explain.
3. What are the factors which affect compaction?
4. What do you mean by wet side of optimum and dry side of optimum? Which side will you consider for field compaction?
5. What is difference between compaction and consolidation of soil?
6. Describe in brief the field applications of compaction test.
7. What are the different methods of compaction of soil in the laboratory?
8. What is the maximum dry density of soil?
9. What will be the dry density of soil at optimum water (moisture) content?
10. Write the field methods of compaction of soil.
11. Explain zero air void line.
12. Compare standard proctor compaction test with modified proctor compaction test.

CHAPTER 6

Laboratory California Bearing Ratio (CBR) Test

6.1 OBJECT

To determine California bearing ratio (CBR) of soil in laboratory.

6.2 THEORY

The CBR test was developed by the California division of highway as a method of classifying and evaluating soil-subgrade and base course material for flexible pavements and in designing base course for airfield pavements. The CBR is defined as the ratio of load corresponding to the chosen penetration to the standard load for same penetration expressed in percentage. The CBR test may be conducted in undisturbed specimen or remolded specimen in the laboratory. In this method a cylindrical plunger of 50 mm diameter is caused to penetrate a pavement component material at 1.25 mm/minute. The loads for 2.5 mm and 5 mm penetration are recorded. This load is expressed as a percentage of standard load. The standard load values are obtained from the Table 6.1.

Table 6.1: Standard load values on crushed stones for different penetration values

Penetration, mm	Standard load, kg
2.5	1370
5.0	2055
7.5	2630
10.0	3180
12.5	3600

6.3 APPARATUS

Refer IS 2720 (Part 16): 1987

- Loading machine
- Compaction rammer
- Perforated plate, tripod and dial gauge
- Annular weight
- Coarse filter paper
- Sieves 4.75 and 20 mm
- Oven

- Balance
- CBR mould, 150 mm diameter and 175 mm high
- Spacer disc
- Penetration plunger: 50 mm diameter
- Surcharge weights
- Rammer 2.6 kg with 310 mm drop and 4.89 kg with 450 mm drop
- Penetration measuring dial gauge
- Straight edge
- Filter paper
- Soaking tank.

6.4 PREPARATION OF SOIL SPECIMEN

Undisturbed Specimen

The undisturbed soil specimens are obtained by fitting a cutting edge to the mould. The mould with cutting edge is pushed into the ground until the mould is full of soil. The soil around the mould is excavated and then the mould with soil is taken out.

The excess soil from top and bottom surface is trimmed with straight edge.

Remoulded Specimen

About 45 kg of dried material is sieved through 20 mm IS sieve. If there is no sufficient material retained on 20 mm sieve, take equal amount of material passing 20 mm sieve and retained on 4.75 mm sieve and then add it to the material which has already passed through 20 mm IS sieve and retained on 4.75 mm sieve. For remoulded soil optimum water content and dry density are required which are determined by IS light compaction or IS heavy compaction. 5.5 kg weight for granular soils and 4.5–5.0 kg for fine grained soil are taken and mixed with water up to the optimum water content or the field water content. The spacer disc is placed at the bottom of the mould over the base plate. A coarse filter paper is placed over the spacer disc. The wet soil is kept in the mould and then compacted either by IS light compaction or by IS heavy compaction. In IS light compaction or standard proctor compaction the soil is compacted in three equal layers by applying 56 blows of 2.6 kg rammer. The blows should be evenly distributed. In IS heavy compaction or the modified proctor compaction, the soil is compacted in five equal layers by applying blows of the 4.89 kg rammer. In static compaction, the correct 'weight of wet soil to obtain the desired density is placed in the mould. The filter paper is placed on the top of soil and then spacer disc is placed on the filter paper. The compaction is attained by pressing in the spacer disk using a compaction machine or jack.

6.5 PROCEDURE

Refer IS 2720 (Part 16): 1987

1. Remove the clamps and lift the mould with compacted soil. Remove the perforated base plate and the spacer disk. Weight the mould with compacted soil.
2. Invert the mould with compacted soil and place it on the filter placed over the base plate. Tighten the clamps of base plate.
3. Place filter paper on the top surface of the sample and place the perforated plate over it.

4. Place the surcharge weights 2.5 or 5.0 kg weight over perforated plate. Keep the mould in water tank for soaking. Let the water enter the specimen both from top and bottom.

5. The tripod and the dial gauge are placed on the top edge of mould to measure swell. Take the initial dial gauge reading. The test set-up is kept in water tank for four days or 96 hours for soaking and then final dial gauge reading is recorded to measure the expension or swelling of the specimen.

6. At the end of soaking, take out the mould and allow it to drain downwards for 15 minutes. Remove the surcharge weight, the perforated top plate and filter paper. Remove the mould with soil from base plate and then weigh the mould with soil. From this weight, water absorbed is obtained.

7. Clamp the mould with specimen over the base plate. Place surcharge weights on specimen sufficient to produce an intensity of loading equal to the weight of the base material (in field) and pavement.

8. To prevent upheaval of soil into the hole of surcharge, place a 2.5 kg annular weight on the soil surface.

9. Place the mould with specimen and base plate under the penetration plunger of the loading machine. In order to make full contact between the surface of specimen and the plunger, place the plunger under a load of about 4 kg.

10. Set the penetration dial gauge and proving ring dial gauge to zero. Apply load to the penetration plunger at the rate of penetration equal to 1.25 mm per minute.

11. Note the load reading at penetration of 0.0, 0.5, 1.0, 1.5, 2.0, 2.5, 3.0, 4.0, 5.0, 7.5, 10 and 12.5.

6.6 PRECAUTIONS

1. Cutting edge must be fitted to the mould while obtaining undisturbed soil sample.
2. The spacer disc must be placed at the bottom of the mould over the base plate.
3. Coarse filter paper must be placed over spacer disc.
4. The blows of the hammer should be evenly distributed.
5. Keep the test set up in water for soaking for four days or 96 hours.
6. Place a 2.5 kg annular weight on soil surface to prevent upheaval of soil into the hole of surcharge.
7. Apply load to the penetration plunger at the rate of penetration equal to 1.25 mm per minute.

6.7 OBSERVATIONS AND CALCULATIONS

The swelling or expansion ratio is calculated as swelling or expansion ratio

$$\frac{H_f - H_i}{H} \times 100\% \tag{6.1}$$

where H_f = Final dial gauge after soaking, mm
H_i = Initial dial gauge reading before soaking, mm
H = Initial height of specimen, mm.

Plot graph between load (in kg) and penetration (in mm).

Figure 6.1 shows a load penetration curve drawn.

Curve is concave upward in the initial portion and hence correction is required. A correction is applied by drawing a tangent to the upper curve at the steepest point on the curve, i.e.

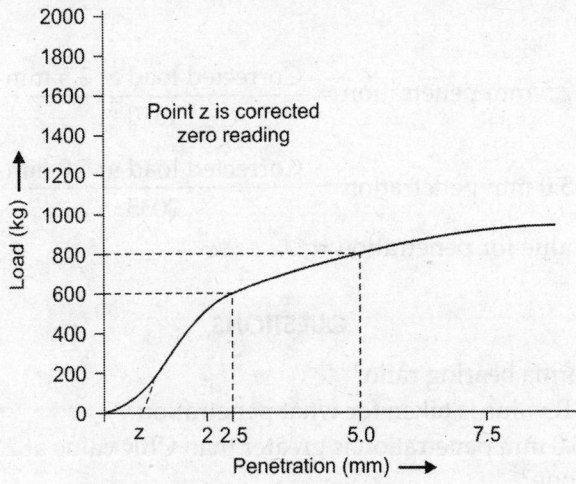

Fig. 6.1: CBR test, load-penetration curve

at point of contraflexure. It intersects the abscissa at point z. This point is the corrected origin and penetration of plunger shall be read from this shifted zero point instead of original zero. The corrected load is read from the corrected penetration value.

The CBR is determined as follows:

$$CBR = \frac{F_c}{F_s} \times 100\% \qquad (6.2)$$

where F_c is corrected test load corresponding to selected penetration from the load penetration curve. F_s is standard load for the same penetration as for corrected load.

The California bearing ratio values are generally calculated at 2.5 mm and 5.0 mm penetration. The CBR value at 2.5 mm penetration is generally greater than at 5.0 mm penetration .

If CBR value corresponding to 5 mm penetration is greater than that for 2.5 mm penetration, the test must be repeated. If same result is obtained then the CBR value corresponding to 5 mm penetration should be taken for design. Table 6.2 shows load penetration values.

Table 6.2: Load penetration values

Penetration (mm)	Proving ring dial gauge reading (divisions)	Load on plunger (kg)	Corrected load (kg)	Standard load (kg)	CBR in %
0.0					
0.5					
1.0					
1.5					
2.0					
2.5					
3.0					
4.0					
5.0					
7.5					
10.0					
12.5					

Result

$$\text{CBR at 2.5 mm penetration} = \frac{\text{Corrected load at 2.5 mm}}{1370} \quad 100\%$$

$$\text{CBR at 5.0 mm penetration} = \frac{\text{Corrected load at 5.0 mm}}{2035} \quad 100\%$$

Hence, final CBR value for penetration = ... %

QUESTIONS

1. What is the California bearing ratio?
2. Generally, the CBR value is taken for what penetration?
3. If CBR value for 5.0 mm penetration is greater than CBR value at 2.5 mm penetration then what should be done?
4. If after repeating the test, the CBR value for 5.0 mm penetration is greater than CBR value at 2.5 mm penetration what should be done?
5. How do you get swell or expansion ratio?
6. What is expression for CBR?
7. Write the applications of CBR.
8. How the correct load is obtained from load - penetration curve?
9. What do you mean by soaking of CBR sample?
10. Why annual load is kept on the sample?

CHAPTER 7

Triaxial Test

7.1 OBJECT

To determine cohesion and angle of shearing resistance for c-Ø soil and stress-strain curve for c-Ø and cohesionless soil.

7.2 THEORY

The triaxial test is the most versatile test to measure the shear strength of soil. In triaxial test a cylindrical specimen is stressed in vertical and lateral directions and the shear parameters cohesion (C) and angle of shear resistance (Ø) are obtained, from which the shear strength of soil is determined. In this test the plane of shear failure is not pre-determined. The triaxial tests are superior where confining stress is to be applied. It is superior than direct shear test. In order to determine cohesion (C) and angle of shearing resistance (Ø) of soil, Mohr's circles are drawn. The tangent line is drawn on the Mohr's circles. It is called the strength envelope. The intercept with ordinate gives cohesion and slope of the line (envelope) gives angle of shearing resistance.

7.3 APPLICATIONS

The C and Ø values obtained from laboratory triaxial test are useful to evaluate the ultimate bearing capacity of soil and also for evaluating the stability of embankment, foundations and slopes. The modulus of elasticity (E) is obtained from triaxial test which is used for designing the flexible pavement in the triaxial method of flexible pavement design.

7.4 TRIAXIAL TEST FOR c-Ø SOIL

7.4.1 Preparation of Specimen

Undisturbed Specimen

Collect the undisturbed sample in a tube of same diameter as the split mould. Transfer the sample to split mould with the help of sample extractor. The sample is then taken out carefully from split mould. If the undisturbed sample obtained from field is of larger diameter, then trim it to desired size.

Remoulded Sample

Prepare the remoulded sample by compacting the soil statically or dynamically at required density and water content in a big size mould. Then trim the sample to desired size, i.e. equal to the diameter of the split mould.

7.4.2 Types of Test

Undrained Test

In this test the outlet valve, i.e. the drainage valve is closed. No drainage is allowed from time of application of lateral pressure (σ_3) till specimen fails. The specimen fails under gradually increasing vertical load.

Consolidated Undrained Test

In this test, the drainage valve is kept open until the sample is fully consolidated under the applied lateral pressure σ_3. After consolidation the drainage valve is closed, and no further drainage is allowed till failure under the application of vertical load.

Consolidated Drained Test

In this test the drainage is allowed during the consolidation of the sample due to lateral pressure σ_3. Then drainage is again allowed under the duration of the application of vertical load.

7.4.3 Procedure (Fig. 7.1)

1. In undrained test put non-porous cap on the bottom pedestal. Put the sample on lower cap then surround the sample by cylindrical rubber membrane which is fixed at lower end, i.e. at pedestal by o-ring. Put non-porous cap on the top of sample and then tie the rubber membrane in the upper cap by o-ring.
2. If in undrained case pore water pressure is to be measured then put the porous stone on the lower pedestal.

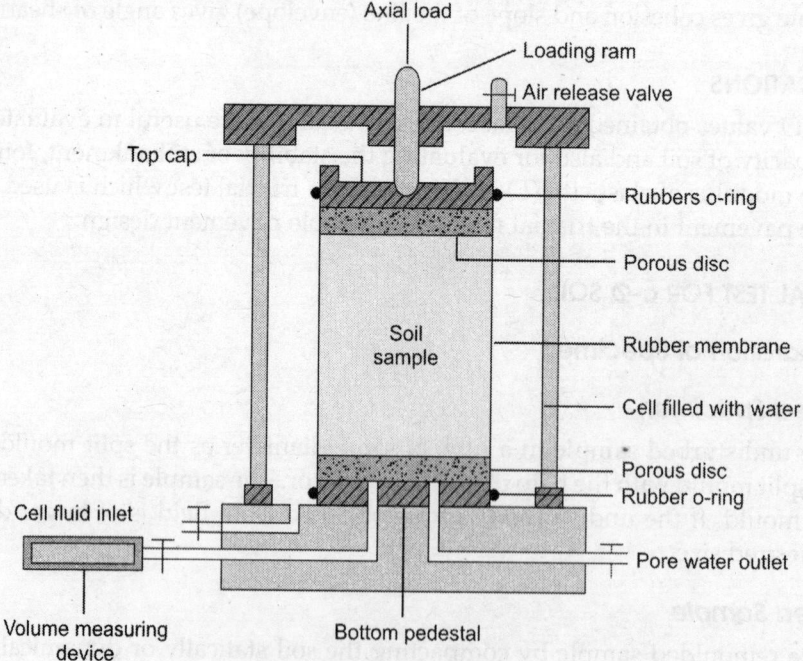

Fig. 7.1: Triaxial apparatus

3. In consolidated undrained and drained test put a porous stone on the lower pedestal, then put sample on it. Then surround the sample by cylindrical rubber membrane which is tied on lower pedestal by o-ring. Tie the sample in top non-porous cap.

4. In order to measure pore water pressure in undrained condition connect the drainage pipe to Bishop's pore pressure apparatus.

5. Apply lateral pressure. Set the proving ring dial gauge and vertical deformation dial gauge to zero. Then apply vertical load and increase gradually until the specimen fails. Take vertical dial gauge reading and proving ring dial gauge reading.

6. Repeat the experiment for various other values of lateral pressure which help in finding the cohesion and angle of internal friction of the soil. Test the soil with lateral pressure of 0, 0.75 and 1.5 kg/cm².

7.5 TRIAXIAL TEST FOR COHESIONLESS SOIL

7.5.1 Dry Sample

I. Preparation of Sample

Put a non-porous cap on the bottom pedestal. Fix the cylindrical rubber membrane at the lower pedestal by o-ring. Put the split mould over the base and the rubber membrane taking it through inside and stretch over it at the top. In order to make a sample of required density, weigh the soil on pan. The soil samples may be in loose dry condition or under dense dry condition. To prepare a sample in loose dry condition, pour the soil inside through funnel. Put non-porous cap on it at the top and then seal it by o-ring. Then take out the split mould. To prepare the sample in dry dense condition, pour the soil in layers and compact each layer by tamping. Then put non-porous cap on it and seal the rubber membrane on it by o-ring. Then take out the split mould.

II. Procedure

1. Put the cylindrical cell properly then fill it with water at required lateral (confining) pressure equal to 0.5 kg/cm².

2. Raise the loading platform such that the loading cap comes in contact with the ram.

3. Set the proving ring (load measuring device) dial gauge and vertical deformation measuring dial gauge to zero.

4. Start the test and take readings of proving ring dial gauge and vertical deformation dial gauge till the sample fails or 20% deformation is achieved.

5. Draw the sketch of the failure pattern of the sample.

6. Repeat the test for fresh sample at the same density as achieved in the first sample for higher cell pressures such as 1.0, 1.5, 2.0, 3.0 and 4.0 kg/cm².

7.5.2 Saturated Sample

I. Preparation of Sample

In order to prepare a sample under saturated condition, put a porous stone (cap) on the pedestal and then seal the rubber membrane to the pedestal by o-ring. Then put the split mould such that the rubber membrane is inside the split mould. Stretch the top portion of the rubber membrane and fold to the split mould. Fill desired water in the membrane. Weigh necessary amount of sand and then pour into the water. When necessary fill sand, level the top surface and then put a non-porous cap on the top and seal it by o-ring. Take out the split mould. Then measure length and diameter of the sample.

II. Procedure

Assemble the cell and then fill it with water. Record the initial readings of burette. Apply confining pressure.

1. Again record the burette readings.
2. Raise the loading platform such that the loading cap comes in contact with the ram.
3. Set the proving ring dial gauge and vertical deformation measuring dial gauge to zero.
4. Start the machine and take the proving ring dial gauge and burette reading and corresponding vertical dial gauge reading till the sample fails or 20% vertical deformation has reached.
5. Draw the sketch of the sample after it has failed.
6. Repeat the test for higher confining pressure for fresh sample at the same density as the first sample.

7.6 OBSERVATIONS AND CALCULATIONS

Correction for area of cross-section.

The normal load applied is the deviatoric load.

The deviatoric stress is load divided by area.

i.e.
$$\sigma_d = \frac{F}{A_0} \, (kg/cm^2) \tag{7.1}$$

But the area of cross-section changes with the deformation of the sample. Hence, correct deviatoric stress is load divided by corrected area.

i.e.
$$\sigma_d = \frac{F}{A_c}$$

$$= \frac{F}{A_0}\left(1 - \frac{\Delta L}{L_0}\right)(kg/cm^2) \tag{7.2}$$

where F is applied load.

A_0 is original area of cross-section (cm^2).

ΔL is the deformation of sample (cm).

L_0 is the original length of specimen (cm).

Hence,   C = ... kg/cm^2

ϕ = ... degrees

All round pressure = σ_3 = ... (kg/cm^2)

The applied vertical stress or deviatoric stress.

= σ_d

Total vertical stress = $\sigma_1 = (\sigma_d + \sigma_3)\,(kg/cm^2)$

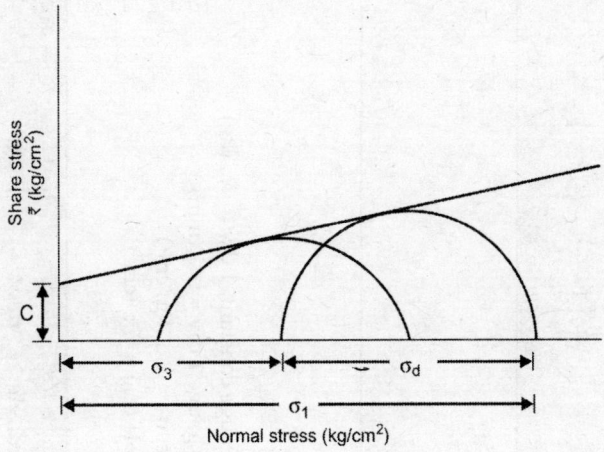

Fig. 7.2: Triaxial text result

Mohr circles arc drawn for various values of σ_3 and corresponding σ_1. The Mohr rupture envelope is then obtained by drawing a tangent to the circles. The intercept of this line with y-axis represents cohesion and the inclination with x-axis represents the angle of internal friction (ϕ) of the soils (Fig. 7.2). Table 7.1 shows triaxial test for c-ϕ soil.

Table 7.1: Triaxial test for c-ϕ soil

Soil type: Type of test: Undrained/consolidated undrained/consolidated drained

Water content = ...%   Proving ring dial gauge constant, 1 div = ... kg

Size of sample = ...cm   Initial length, L_0 = ... cm

Deformation dial gauge constant, 1 div = ...cm   Initial area, A_0 = ... cm^2

S. No.	Lateral pressure σ_3 (kg/cm^2)	Deformation dial reading	Deformation ΔL, (cm)	Strain $\epsilon = \Delta L / L_0$	Proving ring dial reading	Vertical Load (kg)	Deviatoric stress $\sigma_d = F / A_c = \dfrac{F}{A_0}\left(1 - \dfrac{\Delta L}{L}\right)$ (kg/cm^2)	Pore Pressure (kg/cm^2)
1.								
2.								
3.								
4.								
5.								

Table 7.2 shows triaxial test on dry sand.

Table 7.2: Triaxial test on dry sand

Mass of soil taken, Md = ... (g)
Dry unit weight, rd = ... (g/cm³)
Proving ring dial gauge constant, 1 div = ... (kg)
Least count of dial gauge 1 div = ... (mm)

Confining Pressure, σ_3 = ... (kg/cm²)
Length of sample (L_0) = ... (cm)
Diameter of sample (d_0) = ... (cm)
Initial area of sample (A_0) = ... (cm²)

S. No.	Lateral pressure σ_3 (kg/cm²)	Deformation dial reading	Deformation ΔL, (cm)	Strain $\in = \Delta L / L_0$	Proving ring reading	Vertical load F (kg)	Deviatoric stress $(\sigma_1 - \sigma_3) = F/A_c$ $\dfrac{F}{A_0}\left(1 - \dfrac{\Delta L}{L}\right)$ (kg/cm²)	Pore pressure (kg/cm²)
1.								
2.								
3.								
4.								
5.								

Table 7.3 shows triaxial test on saturated sand.

Table 7.3: Triaxial test on saturated sand

Mass of soil taken, Md = ... (g)
Dry unit weight, r_d = ... (g/cm³)
Length of sample (L_0) = ... (cm)
Diameter of sample (d_0) = ... (cm)
Initial volume of sample, V_0 = ... (cm³)

Proving ring dial gauge constant, 1 div = ... (kg)
Least count of dial gauge 1 div = ... (mm)
Confining pressure, σ_3 = ... (kg/cm²)
Initial area of sample (A_0) = ... (cm²)

S. No.	Deformation dial reading	Deformation ΔL, (cm)	Strain $\in = \Delta L / L_0$	Burette reading	Change in volume ΔV (cm³)	New length $L - \Delta L$ (cm)	New volume $V - \Delta V$ (cm³)	Corrected area, Ac $\dfrac{V - \Delta V}{L - \Delta L}$ (cm²)	Proving ring reading	Load, F (kg)	$\sigma_3 = \sigma_3 - \sigma_3$ $= F/Ac$ (kg/cm²)
1.											
2.											
3.											
4.											
5.											

QUESTIONS

1. The triaxial test is most versatile for what?
2. In which directions the cylindrical specimen is stressed?
3. What are the shear parameters?
4. Is the plane of shear failure predetermined?
5. Triaxial test is superior with respect to which test?
6. What is strength envelope?
7. How are the cohesion and angle of shearing resistance determined?
8. What are the applications of triaxial test?
9. How are the undisturbed and remoulded soil sample obtained?
10. What are the different types of triaxial test?
11. How the specimen (soil sample) is arranged in a triaxial test?
12. What apparatus is used to measure pore water pressure?
13. Up to what time the vertical load should be applied?
14. How is the sample prepared for triaxial test for cohesionless soil?
15. What should be the maximum vertical deformation?
16. Up to what time the load is applied?

CHAPTER 8

Plate Load Test

8.1 OBJECT

To determine allowable soil pressure of soil foundation system by vertical plate load test.

8.2 THEORY

In plate load test a test plate, square or circular in shape is used. The plate is placed at the bottom level of the foundation. The plate is then subjected to incremental loading. A load settlement curve is then plotted by measuring the settlement corresponding to each increment of load. The load settlement curve plotted provides the bearing capacity and settlement of the foundation (Fig. 8.1).

8.3 APPLICATIONS

The plate load provides data which is utilised in evaluating allowable soil pressure for shallow foundations. The data obtained from plate load test is also used to determine the coefficient of modulus of subgrade reaction. From plate load test the bearing capacity and settlement of foundation can be determined.

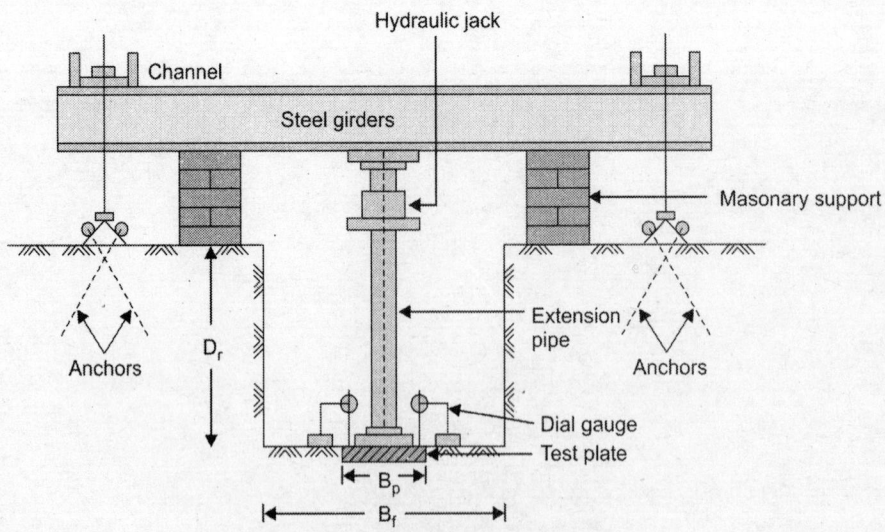

Fig. 8.1: Plate load test set-up

8.4 APPARATUS

- Square or circular mild steel plates of size 30, 45 and 60 cm and thickness 2.5 cm with chaquered or grooved bottom.
- Reaction loading equipment and equipment for gravity loading.
- Remote control type hydraulic jack.
- Proving ring with capacity varying from 10 to 30 tonnes.
- Four dial gauges to measure vertical deflection of 25 to 50 mm range and 0.01 mm sensitivity, datum bars.

8.5 PROCEDURE

1. Mild steel plates having rough bottom and of size 30, 45, 60 cm size and of square shape are used in plate load test.
2. To conduct a plate load test a pit of size $5 B_P \quad 5 B_P$ (where B_P is the size of plate), is excavated up to the depth equal to depth of foundation. A central hole of size $B_P \quad B_P$ is made in the pit. The depth of central hole can be obtained as

$$D_p = \left(\frac{B_p}{B_f} \right) \times D_f \tag{8.1}$$

where B_f is the width of pit, i.e. foundation and D_f the depth of foundation.

3. The plate is seated at the centre over a fine sand layer of maximum thickness equal to 5 mm. The load on the plate is applied through a proving ring and hydraulic jack by taking reaction against a fixed support. The reaction to the jack is transferred by means of a reaction beam, trusses and a loaded platform. The loading on the platform is done by placing sand bags.
4. Initially a seating load equal to 7 kN/m² is first applied then it is released after sometime. The load is then applied in increments of about one-fifth the estimated safe load or one-tenth of the ultimate load up to failure or at least until a settlement of 25 mm has taken place, whichever is earlier.
5. For each load the settlement is recorded after 1, 5, 10, 20, 40, 60 minutes and thereafter at interval of one hour. For clayey soil the observations are continued until the rate of settlement is less than 0.2 mm per hour. Settlements are recorded through four-dial gauges mounted on independent datum and resting on plate.

8.6 LIMITATIONS OF PLATE LOAD TEST

1. The pressure bulb for the plate is at smaller depth than that of the pressure bulb corresponding to foundation. Hence, if soil is not homogeneous and isotropic up to 1½ to 2 B_f the plate load test does not truely represent the actual condition.
2. There is no effect of the size of the plate on the bearing capacity of saturated clay. But the bearing capacity of cohesionless soils increases with the size of plate. To reduce this scale effect it is suggested to repeat the plate load test for two or three different sizes of the plate. Then find the bearing capacity of foundation for each size of plate and then take average of the values obtained.
3. A plate load test is conducted in short duration. It gives total settlement for cohesionless soils. But for clayey soils, it does not give the total (ultimate) settlement.

4. The failure load is well-defined only for general shear failure not for other failures like local shear failure and punching shear failure.

5. As it is not possible to provide reaction more than 250 kN, the plate load test cannot be conducted on a plate larger than 60 cm size of plate.

6. The bearing capacity of soil is affected by the level of water table. Hence, if the water table is above the level of footing, it must be lowered up to the level of foundation.

8.7 OBSERVATIONS AND CALCULATIONS

The load-settlement curve for the test plate is plotted from test data. The load intensity q and settlement plot is made on log-log plot. If the plot shows a break, the pressure corresponding to break is ultimate beating capacity $q_u(p)$. If break is not seen on the plot, the ultimate beating capacity is taken as that corresponding to a settlement one-fifth of the plate width (B_p). If the pressure (q) vs settlement plot is on natural scale, the ultimate bearing capacity is obtained from the intersection of tangents drawn as shown in Fig. 8.2 and Table 8.1.

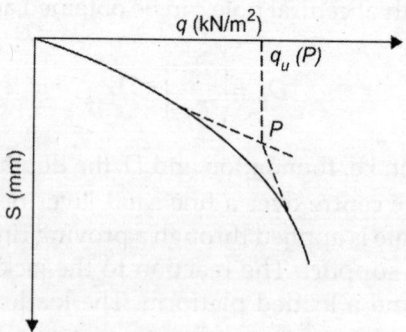

Fig. 8.2: Plate load test result (load settlement curve)

Ultimate Bearing Capacity

(a) For clayey soils

$$q_u(f) = q_u(p) \ (kN/m^2) \tag{8.2}$$

(b) For sandy soils

$$q_u(f) = q_u(p) \ \frac{B_f}{B_p} \ (kM/m^2) \tag{8.3}$$

Settlement

(a) For clayey soil

$$S_f = S_p \times \frac{B_f}{B_p} \ (mm) \tag{8.4}$$

(b) For sandy soil

$$S_f = S_p \times \left[\frac{B_f(B_p + 30)}{B_p(B_f + 30)} \right]^2 \tag{8.5}$$

Where B_f and B_p are in cm.

S_f and S_p are in mm.

Table 8.1: Plate load test

Size of plate = ... (cm)	Date:
Depth of plate = ... (cm)	Name of site:
Size of pit = ... (cm)	Least count of dial gauge=
Depth of pit = ... (cm)	Proving ring constant =

S. No.	Load proving ring reading	Load (kN)	Pressure (kN/m²)	Dial gauge 1		Dial gauge 2		Dial gauge 3		Dial gauge 4	
				R	S (mm)	R	S (mm)	R	S (mm)	R	S (mm)

Where   R = Dial gauge reading

S = Corresponding Settlement

QUESTIONS

1. Write the different sizes of plates used in plate load test.
2. What are the shapes of plates used in plate load test?
3. How will you decide the size of plate to be used in plate load test?
4. What should be the size of the pit, i.e. the foundation?
5. What should be the depth of foundation?
6. Proving ring and hydraulic jack are used for what purpose?
7. How will you decide the load increment in plate load test? What should be the maximum load applied on the plate?
8. Write applications of plate load test.
9. How the allowable bearing pressure of soil-foundation system is obtained in plate load test?
10. How the settlement of foundation is obtained in plate load test?
11. How will you obtain modulus of subgrade reaction from plate load test?

CHAPTER 9

North Dakota Cone Test

9.1 OBJECT

To find out the bearing power of subgrade or *in situ* soil by means of a cone penetrometer of the North Dakota Cone apparatus.

9.2 THEORY

The North Dakota Cone (NDC) test is a penetration test developed by North Dakota State Highway Department of the USA for use in flexible pavement design. This test can be performed *in situ* as well as in laboratory. This equipment being portable and simple can easily be used in field control test of soils, soil-bitumen, etc. for soil which is free from coarse particles. The load carried by the shaft, during penetration into the soil, divided by the area of the cone at the surface level is termed the cone bearing value, q_c.

$$q_c = \frac{Q}{\pi(P_c \tan 7°45')^{\underline{2}}} = \frac{Q}{0.58Pc^2}$$

where q_c = bearing value (kg/cm^2)
            Q = load on cone (kg)
            P_c = Penetration (corrected) of cone

9.3 APPLICATIONS

This test is made use of in flexible pavement design method. An empirical design chart, equation correlating the North Dakota Cone bearing value and pavement thickness have been made available.

9.4 APPARATUS

The North Dakota Cone apparatus consists of a shaft with a sharp cone (angle 15 30′) attached to one end. The movement of the shaft into the soil is measured with the help of a graduated scale. The shaft can be locked or unlocked when necessary. A plate remains fixed to the top of the shaft. Weights can be put on the plate (Fig. 9.1).

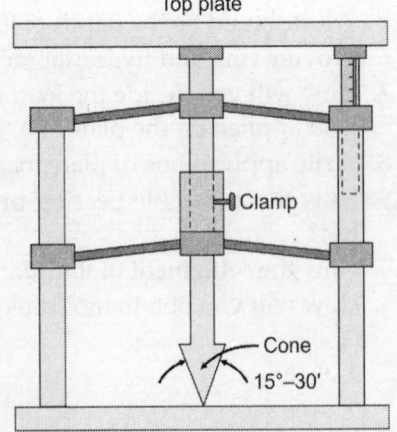

Fig. 9.1: North Dakota Cone test apparatus

9.5 PROCEDURE

1. Level properly the subgrade or *in situ* soil.
2. Keep the apparatus in position and allow the cone to move down such that the tip just touches the surface of soil.
3. Take the initial reading of penetration by locking the shaft.
4. Unlock the shaft and simultaneously start the watch. Allow the penetration for one minute.
5. Lock the shaft and take the reading.
6. Find the difference of penetration readings of step 5 and 3. It gives the penetration of cone under load of 5 kg. The load 5 kg is that of shaft plus cone.
7. Increase the load by 10 kg, i.e. put 5 kg weight on top plate. Note down the penetration reading after unlocking the shaft for one minute.
8. Repeat the procedure for total loads of 20 and 40 kg. This includes the weight of cone and shaft.

9.6 OBSERVATIONS AND CALCULATIONS

If the cone has a true point, for equal bearing pressure, the penetration at 10 kg load should be half that for 40 kg load. Hence correction 'C' due to the rounded point is given by

$$C = P_{40} - 2P_{10}$$

where P_{10} is penetration for 10 kg and
P_{40} is penetration for 40 kg.

The corrected penetration reading.

The corrected penetration reading = Observed penetration reading + correction (C).

Table 9.1 shows bearing value kg/cm^2.

Table 9.1: Bearing value kg/cm^2

Soil type:

Test type:

Correction 'C' = $P_{40} - 2P_{10}$

Load (kg)	Penetration reading P, mm	Corrected penetration, PC, cm	Bearing value Kg/cm^2
0			
5			
10			
20			
40			

Mean NDC Bearing value =

Result

The result is obtained by taking average bearing value excluding the first reading with 5 kg total load. This gives the North Dakota Cone bearing value of the soil.

Limitations

The use of North Dakota Cone test is limited to fine grained soils (silts and clays) free from coarse particles.

QUESTIONS

1. What is the objective of North Dakota Cone test?
2. Who developed North Dakota Cone test?
3. Where the North Dakota test is performed?
4. Define cone bearing value.
5. Write applications of North Dakota Cone test.
6. Write the apparatus for North Dakota Cone test.
7. Write the procedure of performing North Dakota test.
8. What is corrected penetration reading?
9. Write expression for correction due to the rounded point.
10. Write the limitation of use of North Dakota test.

PART 2

Soil Classification

10. Indian Standard Soil Classification

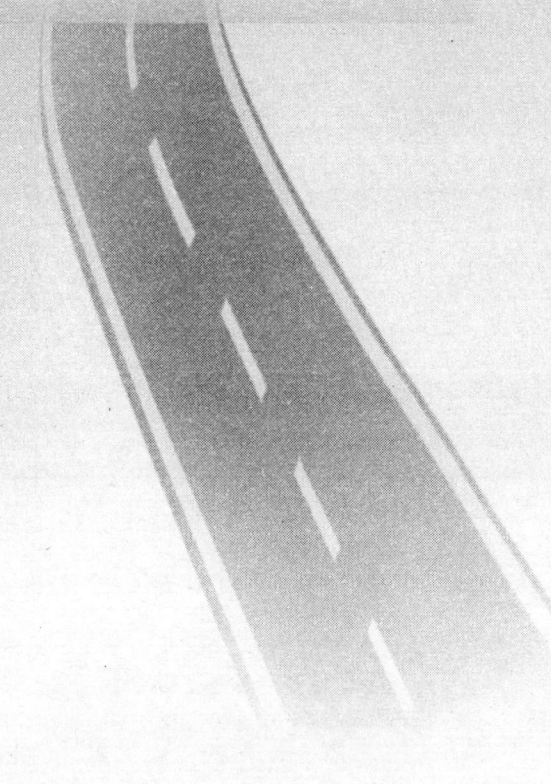

CHAPTER 10

Indian Standard Soil Classification

10.1 OBJECT

Soil classification based on Indian Standard (IS: 1498–1970).

10.2 THEORY

The basic soil components are classified into seven classes as mentioned in Table 10.1 according to Indian Standard (IS: 1498–1970). The soils are mainly divided into two types, viz. coarse grained soils and fine grained soils. Boulders, Cobbles, gravels and sounds are present in coarse grained soil. Silt, clay and organic matter are present in fine grained soil.

Table 10.1: Basic soil components of IS classification (IS: 1498–1970)

S. No. (1)	Soil (2)	Soil components (3)	Symbol (4)	Particle—size range and description (5)
(i)	Coarse-grained	Boulder	None	Rounded to angular, bulky, hard rock particle; average diameter >300 mm
		Cobble	None	Rounded to angular, bulky, hard rock particle; average diameter <300 mm but retained on 80 mm IS sieve
		Gravel	G	Rounded to angular, bulky, hard rock particle; passing 80 mm IS sieve but retained on 4.75 mm IS sieve Coarse: 80–20 mm IS sieve, Fine: 20–4.75 mm IS sieve
		Sand	S	Rounded to angular, bulky, hard rock particle; passing 4.75 mm IS sieve but retained on 75 micron IS sieve Coarse: 4.75–2.0 mm IS sieve Medium : 2.0 mm to 425 micron IS sieve, Fine: 425 to 75 micron IS sieve
(ii)	Fine-grained components	Silt	M	Particles smaller than 75 micron IS sieve; identified by behaviour, that is slightly plastic or non-plastic regardless of moisture and exhibits little or no strength when air dried
		Clay	C	Particle <75 micron IS sieve; identified by behaviour, that is it can be made to exhibit plastic properties within a certain range of moisture and exhibits considerable strength when air dried
		Organic matter	O	Organic matter in various sizes and stages of decomposition

Table 10.2: Soil classification (including field identification and description)

Division	Subdivision	Group letter symbol	Mapping colour	Typical Name	Field Identification Procedures (excluding particles larger than 80 mm and basing fraction on estimated weights)	Information required for describing soils
1	2	3	4	5	6	7
Coarse-grained soils more than half of material is larger than 75 micron IS sieve size. The smallest particle visible to the naked eye. — *Gravels* More than half coarse fraction is >4.75 mm IS sieve size	Clean gravels (little or no fines)	GW	Red	Well graded gravels, gravel-sand mixtures, little or no fines	Wide range in grain sizes and substantial amounts of all intermediate particle sizes	For undisturbed soils add information on stratification; degree of compactness, cementation, moisture conditions and drainage characteristics.
		GP	Red	Poorly graded gravels or gravel-sand mixtures, little or no fines	Predominantly one size or a range of sizes with some intermediate size missing	Give typical name; indicate approximate percentage of sand and gravel maximum size; angularity, surface condition and hardness of the coarse grains; local or geologic name and other pertinent descriptive information; and symbol in parentheses.
	Gravels with (appreciable amount of fines)	GM	Yellow	Silty gravels, poorly graded gravel-sand-silt mixtures	Non-plastic fines or fines with low plasticity (for identification procedures, see ML and MI below)	
		GC	Yellow	Clayey gravels, poorly graded gravel-sand-clay mixture	Plastic fine (for identification procedures, see CL and CI below)	
Sands More than half of coarse fraction is <4.75 mm IS sieve size	Clean sands (little or no fines)	SW	Red	Well graded sands, gravelly sands; little or no fines	Wide range in grain sizes and substantial amounts of all intermediate particle sizes	Example: Silty sand, gravelly; about 20% hard angular gravel particle, 10 mm maximum size; rounded and subangular sand grains; about 15% non-plastic fines with low dry strength; well compacted and moist; in place; alluvial sand (SM)
		SP	Red	Poorly graded sands or gravelly sands; little or no fines	Predominantly one size or a range of sizes with some intermediate sizes missing	
	Sands with fines (appreciable amount of fines)	SM	Yellow	Silty sands, poorly graded sand-silt mixtures	Non-plastic fines or fines with low plasticity (for identification procedures, see ML and MI below)	
		SC	Yellow	Sands with fines (appreciable amount of fines)	Plastic fine (or identification procedures, see CL and CI below)	
(For visual classification the 5 mm size may be used as equivalent to the 4.75 mm IS sieve size)					Identification procedure (on fraction <425 micron IS siever size) — Dry strength / Dilatancy / Toughness	

Contd...

Table 10.2: Soil classification (including field identification and description) *(Contd..)*

Division	Subdivision	Group letter symbol	Mapping colour	Typical Name	Field Identification Procedures (excluding particles larger than 80 mm and basing fraction on estimated weights)			Information required for describing soils
1	2	3	4	5		6		7
Fine-grained soils — More than half of the material is <75 micron IS sieve size — The 75 micron IS sieve size is about the smallest particle visible to the naked eye	Silts and clays with medium compressibility and liquid limit greater than 35 and <50	ML	Blue	Inorganic silts and very fine sand rock flour, silty or clayey fine sands or clayey silts with none to low plasticity	None of low	Quick	None	For undisturbed soils add information on structure, stratification; consistency in undisturbed and remoulded states, moisture and drainage conditions.
		CL	Green	Inorganic clays, gravelly clays, sandy clays, silty clays, lean clays of low plasticity	Medium	None to very slow	Medium	Give typical name; indicate degree and character of plasticity, amount and maximum size of coarse grains; colour in wet condition; odour, if any local or geologic name and other pertinent descriptive information and symbol in parentheses.
		OL	Brown	Organic silts and organic silty clays of low plasticity	Low	Slow	Low	
	Silts and clays with light compressibility and liquid limit >50	MI	Blue	inorganic clays, gravelly clays, sandy clays, silty clayey fine sands or clayey silts of medium plasticity	Low	Quick to slow	None	Example: Clayey silt, brown; slightly plastic; small percentage of fine sand; numerous vertical root holes; firm and dry in place; loess (ML)
		CI	Green	Inorganic clays, gravelly clays sandy clays, silty clays, lean clays of medium plasticity	Medium to high	None	Medium	
		OI	Brown	Organic silts and organic silty calys of medium plasticity	Low to medium slow	Slow	Low	
	Silts and clays compressibility and liquid limit >50	MH	Blue	Inorganic silts of high compressibility, micaceous or diato-maceous fine sandy or silty soils, elastic silts	Low to medium	Slow to none	Low to medium	
		CH	Green	Inorganic clays of high plasticity, fat clays	High to very high none	None	High	
		OH	Brown	Organic clays of medium to high plasticity	Medium to high	None to very slow	Low to medium	
		Pt	Orange	Peat and other highly organic soils with very high compressibility	Readily identified by colour, odour, spongy feel and frequently by fibrous texture			

10.3 TYPES OF SOILS

These are two types:

10.3.1 Coarse Grained Soil

The coarse grained soils are of two types, gravels and sands:

  i. *Gravels (G):* If more than 50% of soil is >4.75 mm IS sieve, the soil is defined as gravel.

  ii. *Sands (S):* If more than 50% of soil is <4.75 IS sieve the soil is defined as sand.

Each of the above subdivisions are further subdivided into following four groups:

W: Well graded

P: Poorly graded

C: Clayey

M: Silty

The types of coarse grained soils can be designated using symbols in combination. For example, SP means poorly graded sand.

Table 10.1 gives symbols and description for basic soil components.

10.3.2 Fine Grained Soil

Fine grained soils are classified on the basis of liquid limit. These soils are as follows:

  i. *Silt and clays of low compressibility:* It has liquid limit less than 35. It is represented by symbol *L*.

  ii. *Silts and clays of medium compressibility:* It has liquid limit >35 and <50. It is represented by symbol *I*.

  iii. *Silts and clays of high compressibility:* It has liquid limit >50. It is represented by symbol *H*.

Table 10.2 gives the field identification, group symbols and typical names for all the soils. The fine grained soils may also be classified with the help of Plasticity Chart shown in Fig. 10.1.

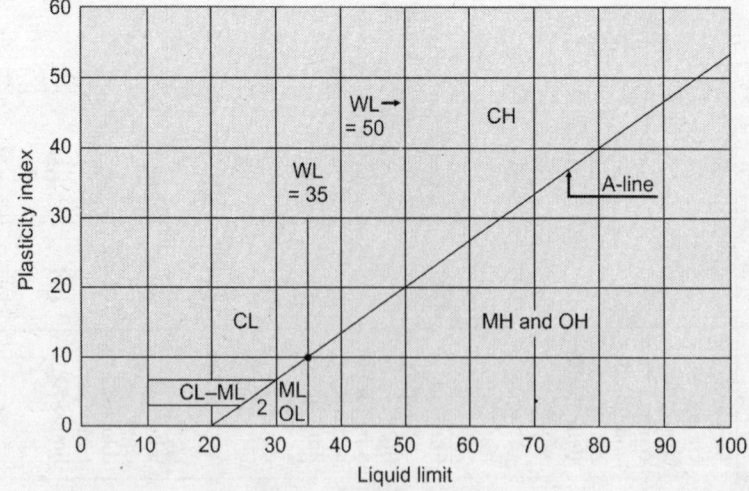

Fig. 10.1: Plasticity Chart

On Plasticity Chart the equation of A-line which divides inorganic clay from silt and organic soil is as follows:

$$I_p = 0.73 \, (W_L - 20)$$

The common boundary classifications for coarse grained soils are GW – SW, GM – GC, etc. The common boundary condition for fine grained soils are ML – MI; CL – CI, etc. The boundary classifications between coarse grained and fine grained soils are: SM – ML and SC – CL.

10.4 PROCEDURE

Do the soil classifications based on Tables 10.1, 10.2 and Plasticity Chart shown in (Fig. 10.1).

10.5 LIMITATIONS

Only Indian standard soil classifications have been discussed. Other classification methods have not been discussed.

QUESTIONS

1. Write soil component and its symbol.
2. Write particle size range and its description.·
3. Write for each soil component, its symbol and particle-size range and description.
4. Define coarse grained soil based on particle size. Is smallest particle visible to naked eye?
5. Define gravel based on particle size.
6. Define sand based on its particle size.
7. Write the group letter symbol and mapping colour for clean gravels with little or no fines.
8. Write the group letter symbol and mapping colour for gravel with appreciable amount of fines.
9. Write the group letter symbol and mapping colour for clean sands with little or no fines.
10. Write the group letter symbol and mapping colour for sands with appreciable amount of fines.
11. Write typical name and field identification procedures for clean gravel with little or no fines.
12. Write the typical name and field identification procedure for gravels with appreciable amount of fines.
13. Write the typical name and field identification procedure for clean sands with little or no fines.
14. Write typical name and field identification of procedure for sand with appreciable amount of fines.
15. Define fine grained soil based on particle size.
16. Write the group letter symbol and mapping colour for silts and clays with medium compressibility and liquid limit >35 and <50.
17. Write the group letter symbol, and mapping colour for silts and clays with light compressibility and liquid limit >50.
18. Write the group letter and mapping colour for silts and clays compressibility and liquid. limit >50.

19. Write the typical name and field identification procedure for silts and clays with medium compressibility and liquid limit >35 and <50.
20. Write the typical name and field identification procedure for silts and clays with light compressibility and liquid limit >50.
21. Write the typical name and field identification procedure for silts and clays compressibility and liquid limit >50.
22. Write group letter symbol, mapping colour, typical name and field identification procedure for highly organic soils.
23. Discuss Plasticity Chart.
24. Write symbol for well graded, poorly graded clayey and silty soils.
25. What is liquid limit and symbol for silts and clays of low compressibility?
26. What is the range of liquid limit and symbol for silts and clays of medium compressibility?
27. What is liquid limit and symbol for silts and clays of high compressibility.
28. Write the equation for A-line.
29. Write the common boundary classifications for coarse grained soils.
30. Write the boundary classifications between coarse grained soil and fine grained soils.

PART 3

Tests on Aggregate

CHAPTER 11

Aggregate Crushing Value Test

11.1 OBJECT

To determine crushing value of given road aggregates.

11.2 THEORY

The aggregate crushing value gives a relative measure of the resistance of an aggregate to crushing under a gradually applied compressive load. Crushing value for aggregate is determined in terms of the percent wear of aggregates on application of load. If the crushing value of aggregate is low, the aggregate will be stronger. If the crushing value of aggregate is more, the aggregate will be weaker. To achieve a high quality of pavement, aggregate possessing low aggregate crushing value should be preferred.

11.3 APPLICATIONS

The aggregate crushing value test can be used to assess the suitability of aggregates with respect to crushing strength for different types of pavement components. The aggregates should be strong enough to withstand high stresses due to wheel loads.

11.4 APPARATUS

The apparatus for the standard aggregate crushing test as per IS 2386: 1963 (Part IV) consists of the followed *cylinder of internal diameter 15.2 cm, square base plate:* Plunger having a piston of diameter 15 cm with a hole provided across the stem of the plunger which facilitate the rod to be inserted so that it can lift and place the plunger in the cylinder.
- *Steel tamping rod:* One end of steel tamping rod is rounded. Its length is 45–60 cm and diameter 1.6 cm.
- *Balance:* It should have capacity of 3 kg with accuracy equal to 1 g.
- *Cylindrical measure:* It has internal diameter equal to 11.5 cm and height equal to 18 cm.
- *IS sieves:* IS sieves of size 12.5, 10 and 2.36 mm.
- *Compression testing machine:* It must apply load of 40 tonnes. The load should be applied at a uniform rate of 4 tonnes per minute (Fig. 11.1).

11.5 PROCEDURE

1. For standard test select the aggregate passing 12.5 mm and retained on 10 mm IS sieves.
2. Put the aggregate in surface dry condition before testing.

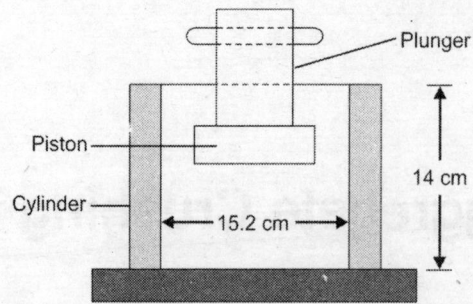

Fig. 11.1: Aggregate crushing test apparatus

3. Dry the aggregate by heating at a temperature of 100 to 110 C for a period of 4 hours and cool it to room temperature.

4. Take about 3.5 kg of sample and then pour the aggregate in the measuring cylinder such that the cylinder is filled more than 1/3rd depth.

5. Tamp the aggregate in the cylinder by 25 number of blows by rounded end of the tamping rod.

6. Similarly, fill the second and third layer in the measuring cylinder, such that it is completely full.

7. Leave off the aggregates at the top of cylindrical measure by tamping rod which acts as a straight edge.

8. Empty the measuring cylinder and weigh the aggregates accurate up to 1 g.

9. Put the aggregate in test mould (cylinder) such that it is filled more than 1/3rd of its height. Tamp it by tamping rod for 25 number of blows.

10. Similarly, put and tamp the second layer and third layer. The total depth of aggregate in the cylinder should be 10 cm after tamping.

11. Level of the surface of aggregates and place the plunger on it so that the plunger rests on levelled surface.

12. Place the cylindrical mould with aggregate and plunger on the pedestal of the compressive machine.

13. Then apply the load of the plunger. The load applied on plunger at a uniform rate of 4 tonnes per minute until the total load is 40 tonnes. Release the load, and remove the test mould.

14. Take out the material from test mould and sieve through 2.36 mm IS sieve.

15. Compute the aggregate crushing value.

16. Repeat the test for second sample. The mean of two observations is reported as the 'Aggregate crushing value'.

11.6 OBSERVATIONS AND CALCULATIONS

Let W_1 = Total weight of dry sample taken.

W_2 = Weight of the portion of crushed material passing 2.36 mm IS sieve.

$$\text{Aggregate crushing value} = \frac{W_2}{W_1} \times 100$$

Table 11.1 shows aggregate crushing value test.

Table 11.1: Aggregate crushing value test

Observation No.	Sample 1	Sample 2
Total weight of dry sample taken = W_1 g		
Weight of portion passing 2.36 mm IS sieve = W_2 g		
Aggregate crushing value = $\dfrac{W_2}{W_1} \times 100$		
Mean crushing value =		

Result

The mean of the crushing value obtained in two tests is reported as the aggregate crushing value.

11.7 DETERMINATION OF TEN PERCENT FINES VALUE

11.7.1 Object

The 10% fines value is a measure of resistance on the aggregates to crushing.

11.7.2 Apparatus

The apparatus and material are same as for the standard aggregate crushing test.

11.7.3 Procedure

1. Place the sample in the cylinder. Place the plunger on levelled aggregates in the cylindrical mould.
2. Place the assembly on the pedestal of compression testing machine.
3. Apply the load for 10 minutes for normal crushed aggregate such that the total penetration of the plunger is 20 mm.
4. Apply the load for 10 minutes to cause a load penetration of plunger of 15 mm for rounded or partially rounded aggregates.
5. Apply the load for 10 minutes to cause a total penetration of 24 mm for honey-combed aggregate.
6. When the maximum penetration is reached, release the load.
7. Take out the aggregate from the cylindrical mould and sieve it on a 2.36 mm IS sieve. Weigh the fines passing 2.36 mm IS sieve.
8. Find the percentage of weight of material passing through 2.36 mm IS sieve with the total aggregate taken. This percentage normally falls in the range of 7.5 to 12.5. If the percentage does not fall in this range repeat the test with necessary adjustment of Load.
9. Carry out two test at load (say W_1 tonnes) such that the percentage fine lies between 7.5 and 12.5. Let the mean of percentage fines be W_2 for calculating the load required for 10% fines.

$$\text{Load for 10 \% fine} = \frac{14W_1}{(W_2 + 4)}$$

11.8 TEST WITH NON-STANDARD SIZE OF AGGREGATES

When non-standard size of aggregates are used for crushing test, the details should be as given on Table 11.2 according to IS 2386: Part IV.

Table 11.2: Details of aggregate crushing test for non-standard sizes of aggregates

Normal Sizes (IS Sieve)		Diameter of cylinder to be used (mm)	Sizes of IS sieve for separating fines (mm)
Passing through (mm)	Retained on (mm)		
25	20	150.0	4.75
20	12.5	150	3.35
10	6.3	150 or 75	1.70
6.3	4.75	150 or 75	1.18
4.75	3.35	150 or 75	.850
3.35	2.36	150 or 75	0.600

11.9 PRECAUTIONS

1. Place the plunger centrally on the aggregate of cylindrical mould such that the plunger does not touch the inside portion of cylinder.
2. The weight of the fines passing through 2.36 mm is sieve plus the weight of aggregate retained on sieve must be equal to the total aggregate taken the difference in the two cases may be of 1 g.
3. While tamping drop the tamping rod gently and uniformly such that it does not touch the inside of the cylindrical mould.

QUESTIONS

1. What is object of aggregate crushing value test?
2. Define aggregate crushing value.
3. Write the applications of aggregate crushing.
4. What are precautions to be taken while conducting aggregate crushing value test?
5. What are the apparatus for standard aggregate crushing test as per IS 2386:1963 (Part IV)?
6. How is the aggregate crushing value non-standard size aggregate evaluated?
7. The aggregate crushing value of material X is 50 and Y is 40. Which one is better and why?
8. Write the procedure of aggregate crushing value test.
9. Write the procedure to find 10% fines value.

CHAPTER 12

Impact Value of Aggregate

12.1 OBJECT

To determine the impact value of the road aggregates.

12.2 THEORY

Toughness is defined as the property of a material to resist impact. The effect of traffic load is to apply impact on the aggregate. Due to impact the aggregates break into smaller pieces. Hence the aggregates must be very tough so that it does not break into smaller pieces due to impact. The aggregate impact value is a measure of the resistance to sudden impact or shock which may differ for some aggregates from its resistance to a slow compressive load.

12.3 APPLICATIONS

The aggregate impact test assess the suitability of aggregate with respect to toughness for use in pavement construction. The maximum permissible aggregate impact values for various pavements are shown in Table 12.1 as recommended by the Indian Roads Congress. The suitability of soft aggregate for road construction has been shown in Table 12.2.

12.4 APPARATUS

The apparatus of the aggregate impact value test as per IS 2386 (Part IV): 1963 consists of following components:

i. *Impact Testing Machine*
   - A testing machine with a metal base with a plane lower surface.
   - A cylindrical steel cup: It has internal diameter equal to 102 mm and depth equal to 50 mm.
   - A metal hammer: It has weight 13.5–14.0 kg. The shape of lower end is cylindrical, having diameter 100 mm and length equal to 50 mm with 2 mm chamfer at the lower edge. The free fall of the hammer is 380 mm.

ii. A cylindrical metal measure: It has internal diameter 75 mm and depth equal to 50 mm. It is used for measuring aggregates.

iii. Tamping rod: It has diameter equal to 10 mm and length equal to 320 mm. One end of the tamping rod is rounded.

iv. **Sieve:** IS sieves of sizes 12.5, 10 and 2.36 mm.

v. **Balance:** Capacity not less than 500 g to weight accurate to 0.1 g.

vi. **Oven:** A thermostatically controlled drying oven capable of maintaining constant temperature between 100 and 110 C (Fig. 12.1).

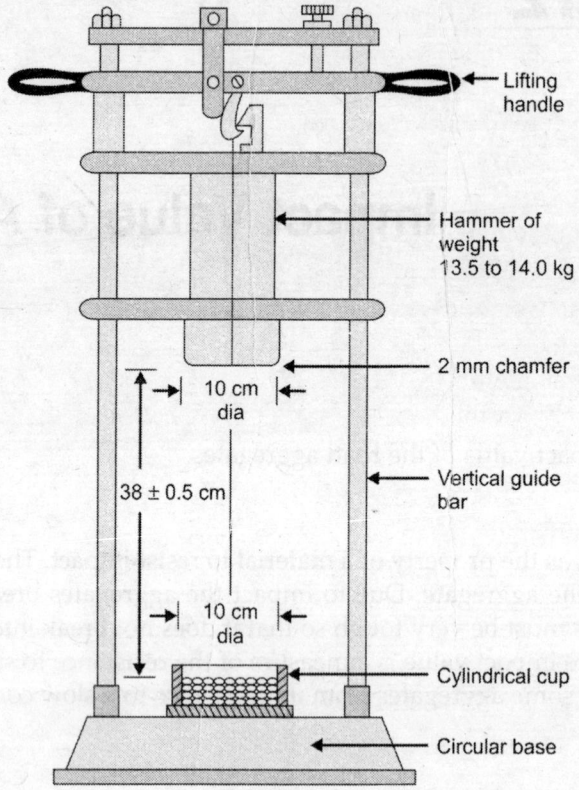

Fig. 12.1: Aggregate impact test set-up

Labels in figure:
- Lifting handle
- Hammer of weight 13.5 to 14.0 kg
- 2 mm chamfer
- 10 cm dia
- 38 ± 0.5 cm
- 10 cm dia
- Vertical guide bar
- Cylindrical cup
- Circular base

12.5 PROCEDURE

1. *Sample:* It consists aggregates passing 12.5 mm sieve and retained on 10 mm sieve. The aggregates should be dried by heating at 100–110 C for a period of 4 hours and than cooled.
2. Fill the cylindrical measure up to 1/3rd of its height by aggregates. Tamp the aggregate by 25 number of blows with rounded end of the tamping rod.
3. Fill the cylindrical measure to 2/3rd of its height and tamp it by tamping rod 25 times. Fill the cylindrical measure such that aggregate overflow. Then tamp it by tamping rod 25 times.
4. Strike off the surplus aggregates.
5. Determine the net weight of the aggregates to the nearest gram. Let it be W_1.
6. Place the impact machine with its bottom plate flat on the floor. It makes the hammer guide column vertical.
7. Fix the cup firmly in position on the base of the machine and transfer the whole of the sample from cylindrical measure to the cup and compact by giving 25 number of blows.
8. Raise the hammer until its lower face is 380 mm above the surface of aggregates in the cup. Allow the hammer to fall freely on the aggregates. Give 15 number of blows at an interval of not less than one second.

9. Remove the crushed aggregates from cup and sieve it through IS 2.36 mm sieve until no further significant amount passes in one minute. Weigh the fraction passing IS 2.36 mm sieve accurate to 0.1 g. Let it be W_2. Also weigh the fraction retained on the sieve.

10. Repeat the above test on fresh aggregate sample.

12.6 PRECAUTIONS

 i. Drop the tamping rod gently and uniformly on the aggregate.

 ii. The tamping rod should not touch the inner side of the mould.

 iii. After sieving the weight of material passing 2.36 mm sieve plus the weight of aggregate retained on sieve must be equal to the total weight of aggregate sample taken.

 iv. Place the plunger such that it falls directly on the sample. It is achieved by placing the plunger centrically. In this case the plunger does not touch the wall of the mould.

12.7 OBSERVATIONS AND CALCULATIONS

Table 12.1 shows aggregate impact value.

Table 12.1: Aggregate impact value

	Sample I	Sample II
Total weight of dry sample taken = W_1 g		
Weight of portion passing 2.36 mm sieve = W_2 g		
Aggregate impact value = $\dfrac{W_2}{W_1} \times 100\%$		
Aggregate impact mean value		

Aggregate impact value is to classify the stones in respect of their toughness property as given in Table 12.2.

Table 12.2: Toughness property

Aggregate impact value	Classification
<10%	Exceptionally strong
10–20%	Strong
10–30%	Satisfactory for road surfacing
>35%	Weak for road surfacing

QUESTIONS

1. What are the apparatus used for aggregate impact test?
2. Describe in brief the procedure of aggregate impact test.
3. What are the precautions to be taken for aggregate impact test?
4. What are the applications of aggregate impact test?
5. Define aggregate impact value and aggregate impact mean value.
6. What do you mean by toughness of the aggregates?
7. Write the aggregates impact value to classify the stones in respect of their toughness property.

CHAPTER 13

Los Angeles Abrasion Values

13.1 OBJECT

To determine the Los Angeles abrasion value.

13.2 THEORY

The aggregate (road stone) used in surface course of highway pavements are subjected to wearing action at top due to movements of traffic. The soil particles present between the wheel and road surface causes abrasion on road aggregates (stones) due to the fast moving traffic fitted with pneumatic tyres and also due to steel tyres of animal drawn vehicles. Hence the road aggregates must be hard enough to resist the abrasion due to traffic.

13.3 APPLICATIONS

Los Angeles Abrasion test is widely accepted as a suitable test to assess the hardness of aggregates used in pavement construction.

13.4 APPARATUS

The apparatus as per IS 2386 (Part IV): 1963 consists of:

- *Los Angeles machine* (Fig. 13.1): It consists of a hollow steel cylinder, closed at both ends. Its internal diameter is 700 mm and length equal to 500 mm. It rotates on a horizontal axis. An opening is provided in the cylinder through which test sample can be put inside the cylinder. A cover is provided on opening which when clamped is dust tight. A removable steel shelf is mounted rigidly on the interior surface of the cylinder.

- *Abrasive charge:* It is cast iron spheres of diameter approximately equal to 48 mm and weight of each ball equal to 390 to 445 g. Consider six to twelve balls for the test.

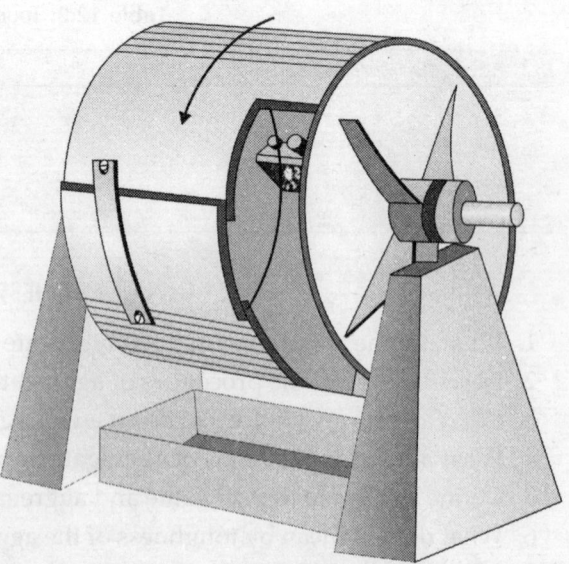

Fig. 13.1: Los Angeles machine

- Sieve: 170 mm IS sieve
- Drying oven
- Balance
- Tray.

13.5 PROCEDURE

1. Take clean aggregate dried in oven at 105–110 C conforming to the grading A to G shown in Table 13.1.

Table 13.1: Grading of test sample

Sieve size (square hole)		Weight in 8 of test samples for grade						
Passing (mm)	Retained (mm)	A	B	C	D	E	F	G
80	63					2500*		
60	50					2500*		
50	40					5000*	5000*	
40	25	1250					5000*	5000*
25	20	1250						5000*
20	12.5	1250	2500					
12.5	10.0	1250	2500					
10.0	6.3			2500				
6.3	4.75			2500				
4.75	2.36				5000			

* Means tolerance ± 2% permitted.

2. Take 5 kg of test specimen for gradings, A, B, C or D and 10 kg for grading E, F, or G and put it in the cylinder.
3. Choose the abrasive charge as per Table 13.2 depending on grading of the aggregate and put into the cylinder of the machine.

Table 13.2: Abrasive charge

Grading	No. of steel balls	Weight of charge, g
A	12	5000 ± 25
B	11	4584 ± 25
C	8	3330 ± 20
D	6	2500 ± 15
E	12	5000 ± 25
F	12	5000 ± 25
G	12	5000 ± 25

4. Fix the cover of cylinder such that it is dust-tight.
5. Rotate the machine at speed of 30 to 33 revolutions per minute.
6. Rotate the machine for 500 revolutions for grading A, B, C and D and 1000 revolutions for gradings E, F and G.

7. Stop the machine after the number of revolutions mentioned in step 6.
8. Take out the material and steel balls from the cylinder. Separate material and steel balls.
9. Sieve the material through 1.70 mm IS sieve.
10. Wash the material coarser than 1.70 mm sieve, dry it in oven at 105–110 C to constant weight and weigh it correct to 1 g.
11. Take another test sample and repeat the test.

13.6 OBSERVATIONS AND CALCULATIONS (Table 13.3)

Table 13.3: Loss Angeles abrasion test

	Sample I	Sample II
Original weight of sample, W_1 = ...g.		
Weight of aggregate retained on 1.70 mm IS sieve, W_2 = ...g		
Loss in weight = $W_1 - W_2$ = ...g		
Wear = $\dfrac{W_1 - W_2}{W_1}$ 100%		
Los Angeles, Abrasion value = Means of two values of sample I and Sample II =		

QUESTIONS

1. How do you express Loss Angeles abrasion value?
2. Write the applications of Los Angeles abrasion test.
3. What precautions will you take while performing Los Angeles abrasion test?
4. Write in brief the procedure for Los Angeles abrasion value test.
5. Write about Los Angeles machines.
6. Why the road aggregates must be hard enough to resist the abrasion due to traffic?
7. Abrasive charge is made of which material?
8. What is the weight of each ball?
9. Why is Los Angeles abrasion test considered superior to other tests?
10. The abrasion values of two aggregates from Los Angeles abrasion test are 50 and 25 respectively. Which aggregate is harder?
11. Why the cover of cylinder of Los Angeles test is kept dust-tight?
12. How many number of balls are considered for the test?
13. The clean aggregate is dried at what temperature?
14. What should be the weight of test specimen?
15. What is the speed of the machine?
16. Write the expression for percent wear of aggregates.

CHAPTER 14

Deval Abrasion Test

14.1 OBJECT

To determine the abrasion value of road aggregates by Deval abrasion test.

14.2 THEORY

In abrasion test, the resistance against wearing is determined as the abrasion value of the aggregates. In Deval abrasion test the test aggregate sample is mixed with six number of standard balls and put inside the cylinder. The inclined cylinder is rotated for 10,000 revolutions at a speed of 30 to 33 rotation per minute. The abrasion value is defined as the ratio of abraded material to the original weight of the sample multiplied by 100 expressed as percentage.

14.3 APPARATUS

The apparatus as per IS 2386 (Part IV): 1963 consists of the following:
- *Deval machine:* The Deval abrasion testing machine consists of one or more cast iron cylinders. Each cylinder has internal diameter 200 mm and depth equal to 340 mm. The cylinder is closed at one end and provided with cover at other end.
- **Abrasion charge:** Six cast iron steel balls are used. The diameter of each ball is approximately 48 mm and weight equal to 390 to 445 g.
- **IS sieve:** IS sieve of size 1.70 mm.
- Balance
- Drying oven and tray (Fig. 14.1).

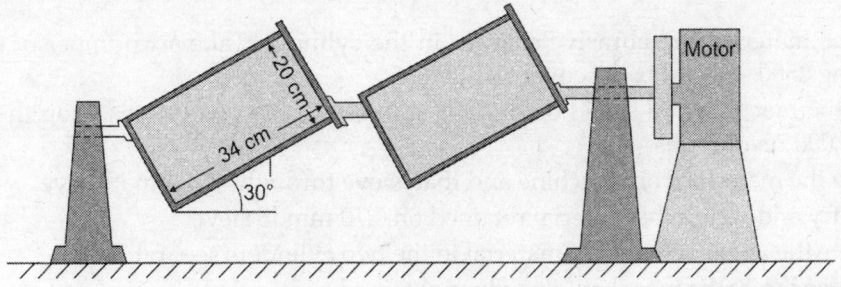

Fig. 14.1: Deval abrasion machine

14.4 PROCEDURE

1. Take the sample of dry and clean aggregates made up of percentage of various sizes conforming to anyone of the gradings given in Table 14.1. Consider the grading which must nearly represents the coarse aggregates to be used in construction. Take the weight of the sample based on the average specific gravity (Table 14.2).

Table 14.1: Gradings of aggregates for Deval abrasion test

Grading	Passing IS sieve (mm)	Retained on IS sieve (mm)	Percentage of sample
A	20	12.5	25
	25	20	25
	40	25	25
	50	40	25
B	20	12.5	25
	25	20	25
	40	25	25
	50	40	25
C	20	12.5	50
	25	20	50
D	12.5	4.15	50
	20	12.5	50
E	10	4.15	50
	12.5	10.0	50

Table 14.2: Weight of sample for Deval abrasion test

Range of specific gravity	Weight of sample 8
Over 2.8	5,500
2.4–2.8	5,000
2.2–2.39	4,500
<2.2	4,000

2. Place the material and abrasive charges in the cylinder. Take six number of steel balls weighing 2500 + 10 g. Fix the cover tightly.
3. Rotate the machine at a speed of 30 to 33 rpm for 10,000 revolutions. Stop the machine after 10,000 revolutions.
4. Remove the material from machine and than sieve through 1.70 mm IS sieve.
5. Wash, dry and weigh the material retained on 1.70 mm IS sieve.
6. Calculate the abrasion value of material in the two cylinders separately.
7. Find the mean of the two abrasion values obtained.

14.5 OBSERVATIONS AND CALCULATIONS (Table 14.3)

Table 14.3: Abrasion value specific gravity

	Sample I	Sample II
Weight of sample, W_1 g		
Weight of material retained on 1.70 mm sieve, W_2 g		
Wear = $\dfrac{W_1 - W_2}{W_1}$ 100%		
Deval abrasion value =		

QUESTIONS

1. What is the objective of Deval abrasion test?
2. Write the expression for abrasion in Deval abrasion test.
3. What are the apparatus used for Deval abrasion test?
4. Write the procedure of Deval abrasion test.
5. How many number of balls are used in the Deval abrasion test?
6. What is the diameter and weight of the balls?
7. Define abrasion.
8. What is the maximum number of revolutions considered?

CHAPTER 15

Specific Gravity and Water Absorption Tests

15.1 OBJECT

To find specific gravity and water absorption for aggregates.

15.2 THEORY

Specific gravity is defined as the ratio of a known *weight of aggregate* to the weight of equal volume of water. Specific gravity indicates the strength or quality of material. If the specific gravity is more, the aggregates will be stronger and if the specific gravity is less, the aggregates will be weaker. Water absorption also indicates the strength of aggregates. If water absorption is more, the aggregates will be weaker and if the water absorption is less the aggregates will be stronger (Table 15.1).

15.3 APPLICATIONS

The specific gravity of aggregates normally used in road construction.

15.4 APPARATUS

- A balance of capacity 3 kg, accurate the 0.5 g.
- Oven to maintain temperature of 100 to 110°C.
- A wire basket of mesh size not >6.3 × 6.3 mm or a perforated container.
- A container for filling water.
- A shallow tray.
- Two dry absorbent clothes, each of size not <750 × 450 mm.

15.5 PROCEDURE

1. Take 2 kg of aggregate sample. Wash it thoroughly to remove fines.
2. Place the sample in the wire basket and immerse it in distilled water. The temperature of water should be between 22 and 32°C. Put water such that it is 50 mm above the top of the basket.
3. Remove the entrapped air from the sample by lifting the basket by 25 mm above the base of tank, by allowing it to drop 25 times. The rate of drop should be one drop per second.
4. Keep the basket and aggregate immersed in water for 24 ± 1/2 hours.

5. Find the weight of basket and aggregate suspended in water. Maintain temperature of 22 to 32 C. Let this weight be W_1.
6. Remove the basket with aggregate from water. Let it drain for few minutes. Transfer the aggregate to one of the dry absorbent clothes.
7. Find the weight of basket suspended in water. Let it be W_2.
8. Surface dry the aggregate by first clothes till its maximum capacity. Then transfer the aggregates in second dry cloth. Spread the aggregate in single layer on the cloth. Dry it at least for 10 to 60 minutes.
9. Do not expose the aggregates to the direct sunlight or any other source of heat.
10. Weigh the surface dried aggregates. Let it be W_3.
11. Place the aggregate in oven in a shallow tray, at temperature of 100 to 110 C for 24 hours.
12. Remove it from oven and cool in an air-tight container. Weigh it. Let it be W_4. It is the dry oven weight of aggregate.

15.6 OBSERVATIONS AND CALCULATIONS

Weight of saturated aggregate suspended in water with basket = W_1 g

Weight of basket suspended in water = W_2 g

Weight of saturated aggregate in water = $(W_1 - W_2) = W_5$ g.

Weight of saturated surface dry aggregate in air = W_3 g.

Weight of water equal to the volume of aggregate = $(W_3 - W_5)$ g.

$$\text{Specific gravity} = \frac{\text{Dry weight of aggregate}}{\text{Weight of equal volume of water excluding air voids in aggregates}}$$

$$= \frac{w_4}{W_3 - W_5} = \frac{w_4}{W_3 - (W_1 - W_2)}$$

$$\text{Apparent specific gravity} = \frac{\text{Dry weight of aggregate}}{\text{Weight of equal volume of water excluding air voids in aggregates}}$$

$$= \frac{w_4}{W_4 - W_5} = \frac{w_4}{W_4 - (W_1 - W_2)}$$

Water absorption = Percent by weight of water absorbed in terms of oven dried weight of aggregates.

$$\frac{W_3 - W_4}{W_4} \times 100\%$$

Table 15.1: Determination of specific gravity and water absorption

(i) Size of aggregate: *Aggregate type =*

Observation No.	Test number		
	1	*2*	*Mean value*
Weight of saturated aggregate and basket in water = W_1 g			
Weight of basket in water = W_2 g			
Weight of saturated surface dry aggregates in air = W_3 g			
Weight of oven dried aggregate in air = W_4 g			
Specific gravity = $\dfrac{W_4}{W_3 - (W_1 - W_2)}$			
Apparent specific gravity = $\dfrac{W_4}{W_4 - (W_1 - W_2)}$			
Water absorption = $\dfrac{(W_3 - W_4)}{W4}$ 100%			

Results

    Mean value of specific gravity =

    Mean value of apparent specific gravity =

    Mean value of water absorption =

15.7 SPECIFIC GRAVITY FOR AGGREGATE <10 MM

15.7.1 Apparatus

- A balance of capacity 3 kg, accurate to 0.5 g.
- Oven to maintain temperature of 100 to 110 C.
- A tray with minimum area 325 cm^2.
- Filter paper
- Funnel
- *Pycnometer:* Pycnometer bottle of about one litre capacity with metal conical cap with 6 mm diameter at top.
- Thermometer capable to measure temperature from 0 to 44 C.

15.7.2 Procedure

1. Take sample of about 1 kg for 10 mm to 4.75 mm or 500 g of finer than 4.75 mm. See that the aggregate is clean and free from any deleterious material.
2. Place screened sample in a tray filled with water. The temperature of water should be 22–32 C. Water should cover at least 50 mm above the top of sample.
3. Remove the air entrapped in or bubbles on the surface of aggregate by slow agitation with a metal rod. Immerse the sample for 24 hours.
4. After immersion spread the aggregate on clean flat surface. Blow warm air on sample until a saturated surface dry condition is achieved.

5. Fill the pycnometer with water. Take the weight of pycnometer and water. Let it be W_1.
6. Put approximately 500 g sample in the pycnometer. Let the weight of pycnometer plus soil be W_2.
7. Fill the remaining part of pycnometer with water. Fix the cap tightly. Let the weight of pycnometer plus soil plus water be W_3.
8. Remove the aggregate from pycnometer. Dry it in oven at temperature of 100 to 110°C for 24 hours . Let the weight of the aggregate be W_4.

15.7.3 Precautions

1. The aggregate should be free from any sticking of air bubble.
2. Weighing should be done carefully.
3. There should be no loss of sample.

15.7.4 Observations and Calculations

$$\text{Specific Gravity} = \frac{W_4}{W_1 - (W_2 - W_3)}$$

QUESTIONS

1. What is the objective of specific gravity and water absorption test?
2. Define specific gravity and apparent specific gravity.
3. Define water absorption.
4. Write the applications of specific gravity.
5. Write the procedure for specific gravity and water absorption for aggregates >10 mm.
6. Write the procedure for specific gravity and water absorption for aggregates <10 mm.
7. What are the precautions for specific gravity and water absorption test for aggregates?
8. Discuss the importance of specific gravity test on road aggregates.

CHAPTER 16

Soundness Test

16.1 OBJECT

To determine soundness of aggregate.

16.2 THEORY

Soundness test is used to determine the resistance to disintegration of aggregate by soaking the aggregate specimen in saturated solution of sodium sulphate or magnesium sulphate to accelerate the effect of weathering into aggregates. The resistance of aggregate to weathering action is defined as soundness of aggregate. Soundness is the property of aggregate to resist excessive changes in volume due to freezing and thawing, thermal changes and alternating wetting and drying. Hence the aggregates must be sound.

16.3 APPLICATIONS

The soundness test is useful to assess the resistance of the aggregate to weathering. According to Indian Roads Congress the maximum permissible loss in soundness test after 5 cycles with sodium sulphate should be 12% for aggregates when the aggregate is used in bituminous surface dressing, penetration macadam and bituminous macadam constructions.

16.4 APPARATUS

Refer IS 2386 (Part V): 1963

- *Containers:* The containers with perforations or with wire mesh.
- *IS sieves:* With square openings and of sizes 4.75, 8.0, 10, 12.5, 16, 20, 25, 31, 40, 50, 63 and 80 mm.
- *Balance:* A balance of capacity 5 kg accurate to 1 g.
- Device for regulating the temperature of the sample.
- *Oven:* Thermostatically controlled maintaining temperature 105 to l 10 C and having an average rate of evaporation of at least 25 g per hour.

16.5 PROCEDURE

Refer IS 2386 (Part V): 1963

1. Prepare the saturated solution of sodium sulphate in water at a temperature of 25 to 30 C. Maintain the solution at temperature of 27 ± 2 C and stirr at frequent interval before it is used. While using the solution, the solution should have specific gravity not <1.151

and not >1.171. Use not less than 420 g of anhydrous salt or 1300 g of crystalline decahydrate salt per litre of water.

2. Alternatively prepare saturated solution of magnesium sulphate in water by dissolving either anhydrous or crystalline magnesium sulphate. While using see that the solution has a specific gravity of not <1.295 and not >1.30 g. Use not <400 g of the anhydrous salt or 1600 g of crystalline heptahydrate per litre of water.

3. Remove the fraction finer than 4.75 mm IS sieve in order to get specimen of coarse aggregate. Keep the sample of such a size that it yields not less than the following amount of the different sizes. Table 16.1 shows aggregate requirement for coarse aggregate.

Table 16.1: Aggregate requirement for coarse aggregate

Sieve Sizes	Quantity required
10–4.75 mm	1000 g
20–10 mm	33%
Consisting of 12.5–10 mm	67%
20–12.5 mm	15000 g
40–20 mm	33%
Consisting of 25–20 mm	67%
40–25 mm	3000 g
63–40 mm	50%
Consisting of 50–40 mm	50%
63–50 mm	Each fraction
80 mm and larger sizes by 20 mm spread in sieve size	3000 g

4. Thoroughly wash the sample (coarse aggregate) and dry it at temperature of 105 to 110 C. Separate the sample to different sizes as mentioned in Table 16.1 by sieving.

5. Weight each fraction and place in separate containers for the test. If the fraction is coarser than 20 mm, count the particles.

6. Immerse the test samples in the sodium sulphate solution or magnesium sulphate solution for 16–18 hours. See that the solution covers the samples to a depth of at least 15 mm.

7. Keep the containers covered to reduce evaporation and during the period of immersion, the temperature of the solution is maintained at 27 ± 1 C.

8. After the immersion period remove the aggregates from solution. Drain it for about 15 minutes.

9. Dry the sample at constant weight at a temperature of 105 to 110 C. Dry the sample to a constant weight at this temperature by checking the weights after 4 hours up to 18 hours. When the successive weights differ by less than 1 g, consider that constant has been attained. Allow it to cool to room temperature.

10. Repeat the immersion, drying and cooling for total of five complete cycles. The number of cycles of alternate immersion and drying are predecided by the experts.

11. Cool and wash the sample such that it is free from sulphate after completion of final cycle. Detect the presence of sodium sulphate in the wash water by reaction of water with barium chloride.

12. Dry each fraction of sample at constant temperature of 105 to 110°C and than allow it to cool at room temperature and then weigh it. Coarse aggregate fractions are sieved by IS sieves of sizes given in Table 16.2.

Table 16.2: Size of aggregate and sieve

Size of aggregate	Sieve size used to determine loss
63 to 40 mm	31.5 mm
40 to 20 mm	16.0 mm
20 to 10 mm	8.0 mm
10 to 4.75 mm	4.0 mm

13. Examine visually each fraction of aggregate to see if there is any evidence of excessive splitting, crumbling or disintegration of grains.
14. Perform combined sieve analysis by combining all sizes on set of sieves of 150 mm, 300 micro meter, 600 micro meter, 1.18, 2.36, 4.75, 10, 20 40 and 80 mm. Note the variation of grain size distribution from this sieve analysis with the original grain size distribution of the sample.

16.6 PRECAUTIONS

1. The aggregate must be clean and dry before first cycle of test.
2. Cool the aggregate to room temperature and then place it in the solution.
3. Sample should be covered while drying.
4. Put heat resistance globes in hand while removing samples from oven.
5. Be careful of barium chloride as it is poison

16.7 OBSERVATIONS, CALCULATIONS AND RESULTS

The results should be repotted giving the following particulars:

1. Type of solution used for test and the number of cycle for the test.
2. Weight of each fraction of sample before test.
3. Calculate percentage loss for each test portion as follows:

$$\text{Percentage Loss} = \frac{W_0 - W_R}{W_0} \times 100$$

where W_0 = Original weight of the test sample

W_R = Weight of aggregate retained on sieve after the sample has been tested.

4. Compute the weighted average loss of coarse aggregate from the percentage of loss for each fraction.
5. For particles size coarser than 20 mm, the number of particles in each fraction before test and the number of particles affected classified as the number disintegrating, splitting, crumbling, cracking flaking, etc. shall be reported. Table 16.3 shows soundness test.

Table 16.3: Soundness test

Type of reagent used:

Type of coarse aggregate sample: *Number of cycles:*

Sieve size mm		Grading of original sample %	Weight of test fraction before test, g	Percentage passing finer sieve after test (actual % loss)	Weight average (corrected % loss)
Passing	*Retained*				
1	2	3	4	5	$6 = \dfrac{5 \times 3}{100}$
60	40				
40	20				
20	10				
10	4.75				

Number of particles coarser than 20 mm before test			*Number of particles affected, classified as the number disintegrating, splitting, crumbling, cracking or flaking*
Passing	*Retained*	*Number before test*	
40 mm	20 mm		
60 mm	40 mm		

QUESTIONS

1. What for saturated solution of sodium sulphate and magnesium sulphate are taken?
2. Write the applications of soundness test.
3. What apparatus are used for soundness test?
4. Write the procedure of soundness test.
5. What are the precautions to be taken for soundness test?
6. Define soundness.
7. How is soundness test value expressed?
8. Which weathering effects are simulated by soundness test?
9. What are the salts usually used in the soundness test for aggregates?
10. What do you mean by soundness test for aggregate?

CHAPTER 17

Shape Test

17.1 OBJECT

To determine the shape of aggregate.

17.2 THEORY

The particle shape test gives information about the presence of flaky, elongated and angular aggregate. The presence of flaky and elongated aggregates are undesirable as they cause inherent weakness leading to breaking down under compaction or heavy structural loads. The angular shape of particles are good as granular base course. This is due to the fact that there is stability due to the better interlocking of aggregates. Thus evaluation of shape of the particles, particularly with reference to flakiness, elongation and angularity is necessary.

17.3 APPLICATIONS

In pavement construction flaky and elongated particles should be avoided. Indian Road Congress has recommended the maximum allowable limits of flakiness index for various types of construction. There is no specified limits of elongation index value.

It has been found that higher is the angularity number more angular and less workable is the aggregate mix. In flexible pavement construction methods angular aggregates are preferred due to higher stability caused by interlocking and friction.

17.4 FLAKINESS INDEX

17.4.1 Definition

The flakiness index of aggregates is the percentages by weight of the particles whose least dimension (thickness) is less than 0.6 of their mean dimension.

17.4.2 Apparatus
- Standard thickness gauge.
- IS sieves of sizes 63, 50, 40, 31.5, 25, 20, 16, 12.5, 10 and 6.3 mm.
- Balance (Fig. 17.1).

17.4.3 Procedure
1. Sieve the sample with sieves as mentioned in Table 17.1.
2. Take a minimum of 200 pieces of each fraction and weight it. Let it be W_1.

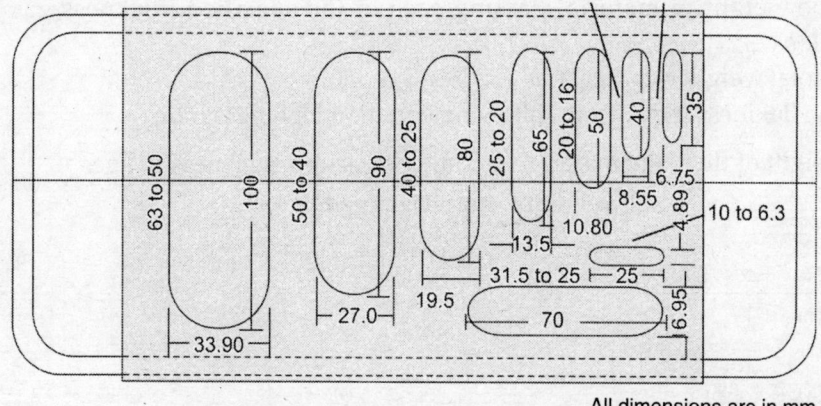

All dimensions are in mm

Fig. 17.1: Flakiness measurement gauge

3. Gauge each fraction for thickness on a thickness gauge to separate flaky material.
4. Weigh the amount of flaky material passing the gauge to an accuracy of at least 0.1% of the weight of the sample.
5. Calculate the thickness of thickness gauge given in Tables 17.1 and 17.2.

Table 17.1: Dimensions of thickness gauge

Size of aggregate		Thickness gauge
Passing through IS sieve mm	Retained on IS sieve mm	(0.6 times the mean sieve) mm
63	50	33.90
50	40	27.0
40	25	19.5
31.5	25	16.95
25	20	13.50
20	16	10.80
16	12.5	8.55
12.5	10.2	6.75
10	6.3	4.89

Let 200 pieces of aggregate pass 50 mm sieve and retained on 40 mm sieve. The width of the appropriate gauge of the thickness gauge is

$$= \frac{50 + 40}{2} \times 0.6 = 27.0 \text{ mm}$$

6. Try to pass each particle (aggregate) through the width, i.e. thickness of the thickness gauge.
7. Weigh the fraction of aggregate passing and retained on specified sieve. Let it be W_1, W_2, W_3, W_4, ... Find the total weight as
$$W = W_1 + W_2 + W_3 + W_4 + ...$$

8. Find the weight of material passing each of the specified thickness gauge. Let it be $w_1, w_2, w_3, w_4, ...$

   Hence total weight $= w = w_1 + w_2 + w_3 + w_4$

9. Find the flakiness index from following equation Flakiness Index

$$\frac{\text{Total weight of the flaky material passing the various thickness gauges}}{\text{Total weight of sample gauged}} \times 100$$

$$= \left(\frac{w_1 + w_2 + w_3 + w_4 + ...}{W_1 + W_2 + W_3 + W_4 + ...} \right) \times 100$$

$$= \frac{w}{W} \times 100\%$$

17.4.4 Observations and Calculations

Table 17.2: Flakiness index

Size of aggregate		Weight of the fraction consisting of at least 200 pieces	Thickness gauge size, mm	Weight of aggregates in each fraction passing thickness gauge, g
Passing through IS sieve mm	Retained on IS sieve mm			
63	50	$W_1 =$	23.90	$w_1 =$
50	40	$W_2 =$	27.0	$w_2 =$
40	25	$W_3 =$	19.50	$w_3 =$
31.5	25	$W_4 =$	16.95	$w_4 =$
25	20	$W_5 =$	13.50	$w_5 =$
20	16	$W_6 =$	10.80	$w_6 =$
16	12.5	$W_7 =$	8.55	$w_7 =$
12.5	10.0	$W_8 =$	6.75	$w_8 =$
10.0	6.3	$W_9 =$	4.89	$w_9 =$

Flakiness Index

$$= \left(\frac{w_1 + w_2 + w_3 + w_4 + ...}{W_1 + W_2 + W_3 + W_4 + ...} \right) \times 100$$

$$= \frac{w}{W} \times 100\%$$

17.5 ELONGATION INDEX

17.5.1 Definition

The elongation index of an aggregate is the percentage by weight of particles whose greatest dimension (length) is greater than 1.8 times their mean dimension.

17.5.2 Apparatus

- Length gauge shown in Fig. 17.2

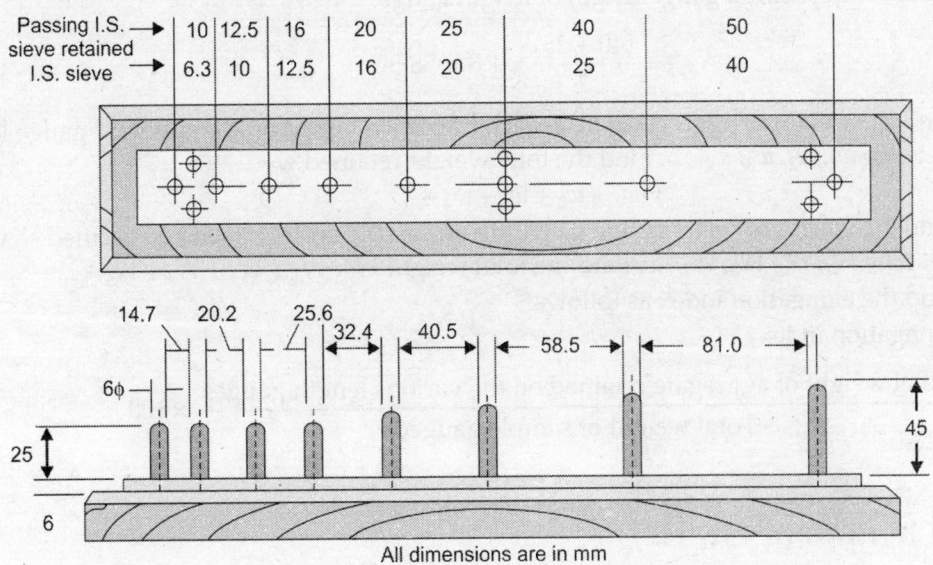

All dimensions are in mm

Fig. 17.2: Elongation measurement gauge

- **Sieves of sizes specified in** Table 17.3.

Table 17.3: Dimensions of length gauge

Size of aggregate		Length gauge
Passing through IS Sieve, mm	Retained on IS sieve, mm	(1.8 times mean sieve), mm
63	50	–
50	40	81.0
40	25	58.5
31.5	25	–
25	20	40.5
20	16	32.4
16	12.5	25.6
12.5	10.0	20.2
10.0	6.3	14.7

17.5.3 Procedure

1. Sieve the sample through IS sieves (Table 17.4).
2. Take minimum 200 pieces of each fraction and then take its weight.
3. To get elongated material gauge each fraction individually for length in a length gauge (Fig. 17.2) and (Table 17.4).
4. Test pieces of aggregates of each fraction through specified gauge length and find the total weight of aggregate retained on length gauge accurate to at least 0.1% of the weight of the test sample.

5. Find the weight of aggregates retained on the specified gauge length of length gauge. Consider the aggregate having 200 pieces passing 40 mm sieve and retained on 25 mm sieve. The specified gauge length of length gauge.

$$= \frac{(40 + 25)}{2} \times 1.8 = 58.5$$

6. Find the weight of aggregates from each fraction retained on the specified gauge length. Let it be $w_1, w_2, w_3, w_4, \ldots$. Find the total weight retained w

$$= w_1 + w_2 + w_3 + w_4 + \ldots$$

Find the weight of each fraction of aggregate passing and retained on specified sieve sizes. Let it be $W_1, W_2, W_3, W_4, \ldots$. Find the total weight $W = W_1 + W_2 + W_3 + W_4 + \ldots$

Find the elongation index as follows:

Elongation index

$$\frac{\text{Total weight of aggregate retained on the various length gauges}}{\text{Total weight of sample gauged}} \times 100$$

$$= \left(\frac{w_1 + w_2 + w_3 + w_4 + \ldots}{W_1 + W_2 + W_3 + W_4 + \ldots} \right) \times 100$$

$$= \frac{100 \, w}{W}$$

17.5.4 Observations and Calculations

Table 17.4: Elongation Index

Size of aggregate		Weight of the fraction consisting of at least 200 pieces g	Length gauge size, mm	Weight of aggregates in each fraction retained on length gauge, g
Passing through IS sieve mm	Retained on IS sieve mm			
63	50	$W_1 =$	–	–
50	40	$W_2 =$	81.0	$w_1 =$
40	25	$W_3 =$	58.0	$w_2 =$
31.5	25	$W_4 =$	–	–
25	20	$W_5 =$	4.5	$w_3 =$
20	16	$W_6 =$	32.4	$w_4 =$
16	12.5	$W_7 =$	25.5	$w_5 =$
12.5	10.0	$W_8 =$	20.2	$w_6 =$
10.0	6.3	$W_9 =$	14.7	$w_7 =$
Total		$W =$	–	$w =$

Elongation Index

$$= \left(\frac{w_1 + w_2 + w_3 + w_4 + \ldots}{W_1 + W_2 + W_3 + W_4 + \ldots} \right) \times 100$$

$$= \frac{w}{W} \times 100\%$$

17.6 ANGULARITY NUMBER

17.6.1 Object

To find angularity number of aggregate.

17.6.2 Definition

The angularity number of an aggregate is the amount by which the percentage voids exceed 33 after being compacted in a prescribed manner.

17.6.3 Theory

By shape and angularity of the aggregates the degree of packing of same size aggregates is obtained when spherical particles are packed densely it has total solid volume equal to 67%. The volume of voids is 33% of total volume. Angular shape particles when packed densely has reduced volume and increase in the volume of voids. The angularity number of an aggregate is the amount by which the percentage voids exceed 33.

17.6.4 Apparatus

• **A metal cylinder** closed at one end and of about 3 litre capacity. The diameter and height each is equal to 156.4 mm.
• A metal scoop of about one litre capacity and size 200   100   50 mm.
• A metal tamping rod of circular cross-section, 16 mm diameter and 600 mm length, rounded at one end.
• A balance of capacity 10 kg to weight up to 1.0 g.

17.6.5 Procedure

1. Determine the weight of water at 27 C required to fill the cylinder.
2. Sieve the aggregate sample in the pair of sieves shown in Table 17.5, such that after separation at least 10 kg of aggregate sample becomes available. Dry the sample in oven at 100–110 C for 24 hours, then cool in air before commencing the experiment.

Table 17.5: Order of sieves for angularity number test

Sample	Passing through sieve, mm	Retained on sieve, mm
1	20	16
2	16	12.5
3	12.5	10
4	10	6.3
5	6.3	4.75

3. The cylinder of 3 litre capacity is suitable for all aggregate samples as mentioned in Table 17.6. If the aggregate is of size >20 mm, the volume of cylinder should be >3 litres and if the aggregate is of size < 4.75 mm, a smaller cylinder may be used.
4. Fill the cylinder up to height slightly greater than one-third of height of cylinder. Compact it by 100 number of blows of tamping rod. The tamping rod should fall freely from height 5 cm above the aggregate surface. Similarly fill the second and third layer as the first layer.

5. After compacting third layer fill the cylinder by aggregates to over flowing. Then struck off the excess aggregates by tamping rod as straight edge.

6. Then add individual pieces of aggregates one by one so as to fill the cylinder up to top. Tamp and roll the rod by placing it horizontally on cylinder edge. Repeat the process until no more aggregates can be added. Take care that no aggregate project above the plane forming the horizontal surface.

7. Weigh the cylinder to the nearest 5 g with the compacted aggregate in it. Repeat the test three times and obtain the mean weight of the aggregate in the cylinder by subtracting the weight of the cylinder.

8. If the result of anyone of the tests differs from mean value by more than 25 g, three more tests are to be immediately performed and the mean of all six tests is calculated.

17.6.6 Observations and Calculations

Table 17.6: Angularity number

Particulars	Observation number							
	1	2	3	Mean	4	5	6	Mean
Weight of aggregate filling the cylinder, g				W_1				W_1

The angularity number is calculated from the formula:

$$\text{Angularity number} = 67 - \frac{W_1}{W_2 G}$$

where
W_1 = Mean weight of aggregates in the cylinder, g
W_2 = Weight of water required to fill the cylinder, g
G = Specific gravity of aggregate.

Express the angularity number to nearest whole number.

QUESTIONS

1. What are the different shapes of aggregates?
2. Write the applications of shape test.
3. What are the apparatus used for flakiness index?
4. What are the apparatus used for elongation index?
5. What are the apparatus used for angularity number?
6. Write the procedure to determine flakiness index.
7. Write the procedure to determine elongation index.
8. Write the procedure to determine angularity number.
9. Define flakiness index, elongation index and angularity number.
10. Write expressions for flakiness index, elongation index and angularity number.
11. What is the significance of shape of aggregate in pavement construction?
12. Discuss the effects of flaky and elongated aggregate in road construction.
13. What parameter do you measure by the shape test?

CHAPTER 18

Stripping Value of Road Aggregate

18.1 OBJECT

To determine the stripping value of aggregates used in road construction.

18.2 THEORY

If water is not present there is no adhesion problem between bitumen and aggregate. If water is present it will cause two problems. First, when aggregate is wet and cold, it is not possible to coat the aggregate with bituminous binder. Secondly, there is stripping of coated binder from the aggregate (Table 18.1). The stripping is caused due to the fact that aggregates have, greater affinity towards water than bituminous binder. There are several laboratory tests to determine the adhesion of bitumen binder to aggregate. These tests are as follow:

1. Static immersion tests
2. Dynamic immersion tests
3. Chemical immersion tests
4. Immersion mechanical tests
5. Immersion trafficking tests
6. Coating tests
7. Adhesion tests, etc.

18.3 APPLICATIONS

Simple static immersion test or the stripping test would be suitable to assess whether the binder would adhere to the aggregate when immersed in water. Indian Roads. Congress has specified the maximum stripping value as 25% for aggregates which will be used in bituminous construction. The examples are surface dressing penetration macadam, bituminous macadam and carpet.

18.4 APPARATUS

- Thermostatically controlled water bath
- Beaker
- Mixer, etc.

18.5 PROCEDURE

Refer IS 6241 : 1571

1. Take approximately 200 g aggregate passing 20 mm IS sieve and retained on 12.5 mm IS sieve.
2. Dry, clean and mix 5% of bitumen binder by weight of aggregate and then heat the binder at 160 C.
3. Heat the aggregates to a temperature of 100 C and 150 C when it is to be mixed with for tar and bitumen respectively.
4. Mix the binder and aggregate thoroughly till the aggregates are completely coated.
5. After complete coating, transfer the mixture to a 500 ml beaker. Then allow it to cool at room temperature for about 2 hours.
6. Add distilled water to immerse coated aggregates.
7. Cover the beaker and keep in a water bath maintained at 40 C for a period of 24 hours.
8. After 24 hours take out the beaker and cool it at room temperature. Estimate the extent of stripping visually while the specimen is still under water. The stripping value is the ratio of the uncovered area observed visually to the total area of aggregates in each test, expressed as percentage.
9. Report the mean of three results as stripping value of tested aggregates and express it as the nearest whole number.

18.6 OBSERVATIONS AND CALCULATIONS

Table 18.1: Stripping test

Type of aggregate: Total weight of aggregate =
Type of binder: Total weight of binder =
Percentage of binder used :   Temperature of water bath =

Sl No.	Stripping percentage	Average value
1.		
2.		
3.		

QUESTIONS

1. What is the objective of stripping value of road aggregate?
2. Write the applications of stripping value test.
3. Write the apparatus used for stripping value test.
4. Write the procedure for stripping value of road aggregates.
5. Define stripping value of road aggregates.
6. What is the significance of stripping value test?
7. Explain stripping.

PART 4

Tests on Bituminous Materials

CHAPTER 19

Softening Point of Bituminous Material

19.1 OBJECT

To determine softening point of bitumen/tar.

19.2 THEORY

According to IS 334: 1982 softening point is the temperature in 0 C at which a standard ball passes through a sample of bitumen in a mould (ring) and falls through a height of 250 mm when heated under water or glycerin at specified condition of test.

The softening point can also be defined as temperature at which the bitumen or tar attains a particular degree of softening. The widely used procedure in road construction is to liquefy the bitumen by heating. The most direct, safest and most common method of determining the temperature susceptibility is the ring and ball softening point test.

19.3 APPLICATIONS

From softening point we get an idea of temperature at which the bituminous material attains a certain viscosity. In warmer place bitumen with higher softening point is preferred. Softening point is also sometimes used to specify hard bitumen and pitches.

19.4 APPARATUS

Refer IS 1205: 1978
1. The ring and ball apparatus consists of the following:
    i. *Steel balls:* Two number, each of diameter 9.5 mm and weight 2.5 ± 0.5 g.
    ii. *Brass rings:* Two numbers each with depth equal to 6.4 mm. Inside diameter to top and bottom are 17.5 mm and 15.5 mm. Outside diameter is 20.6 mm.
    iii. Ball guides to guide the movement of steel balls centrally.
    iv. *Support:* The support holds ring in position. It also allows the suspension of thermometer.
2. **Thermometer:** Which can read up to 100 C with an accuracy of 0.2 C.
3. **Bath:** A heat resistance glass beaker of 85 mm diameter and 120 mm depth.
4. **Stirrer:** Mechanical stiffer shall be used for uniform distribution of heat.

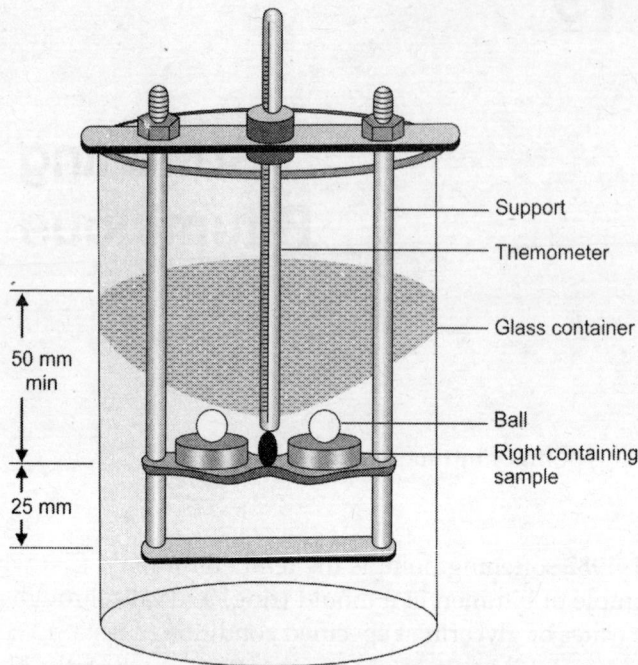

Fig. 19.1: Ring and ball apparatus

19.5 PROCEDURE

Refer IS 1205: 1978

1. Assemble the ring and ball apparatus with rings and ball guides in position (Fig. 19.1).
2. Heat the bitumen between 75 and 100°C above the approximate softening point. The bitumen get converted to fluid completely. Pour it in heated rings placed on metal plate.
3. Cool the rings in air for 30 minutes. Trim the excess bitumen. Put the balls on bitumen on ring.
4. Fill the beaker with water/glycerine to a height of approximately 50 mm above the upper surface of the rings. Use glycerine if the expected softening point is more than 80°C.
5. Heat the ring and ball apparatus such that the temperature of water rises by 5°C per minute. Due to heating bitumen softens and the balls pass through rings and touches the bottom plate by sinking.
6. The temperature at the instant when each of the balls and sample (bitumen) touches the bottom plate of support is recorded as softening point value.
7. Make at least two observations. The mean of two observations (temperatures) recorded for two samples is reported as softening point.

19.6 PRECAUTIONS

1. Use distilled water in the test.
2. Take care that there is no vibration in the apparatus while performing the test.
3. Put the thermometer such that its bulb is approximately at the same level as the rings.

19.7 OBSERVATIONS AND CALCULATIONS

Approximate softening point =
Liquid used in bath = water/glycerine
Duration of air cooling, minutes =
Duration of cooling in water bath, minutes =
Table 19.1 shows rate of heating.

Table 19.1: Rate of heating

Time minutes	Temperature C	Time, minutes	Temperature C
0		11	
1		12	
2		13	
3		14	
4		15	
5		16	
6		17	
7		18	
8		19	
9		20	
10			

Table 19.2 shows observations.

Table 19.2: Observations

Test property	Temperature at which sample touches bottom plate	Reputability	Reproducibility
Sample No.1	Ball No.1		
	Ball No.2		
Sample No. 2	Ball No.1		
	Ball No.2		
Mean value of softening point			

QUESTIONS

1. Define softening point of bitumen.
2. Write the applications of softening point test.
3. Write the apparatus used for softening point test.
4. Write the procedure of softening point test.
5. What precautions will you taken for softening point test?
6. What is softening point?
7. What does softening point of bituminous material indicate?
8. What are the applications of ring and ball test results?
9. What precautions will you take while performing the softening point test?
10. What is softening point?
11. What is the use of softening point in highway construction?
12. What are the factors which affect the softening point test results?

CHAPTER 20

Ductility Test

20.1 OBJECT

To find the ductility of bituminous material.

20.2 THEORY

The bituminous binders should form ductile thin film around the aggregates. This improves the physical interlocking of the aggregates. If the binder material does not possess sufficient ductility, it will crack causing pervious pavement surface. The ductility of a bituminous material is measured by the distance in centimeters to which it will elongate before breaking when two ends of standard briquette specimen of the material are pulled apart at a specified speed and specified temperature. The test is performed at temperature 27 ± 0.5 C and rate of pull of 50 ± 2.5 mm per minute.

20.3 APPLICATIONS

A bitumen binder should have certain minimum ductility. A bitumen having low ductile value when used in bituminous pavement, the pavement may crack. A bitumen has ductility value from 5 to over 100. Often a minimum ductility value of 50 cm is specified for bituminous construction.

20.4 APPARATUS

The apparatus for the standard ductility test as per IS 1208: 1978 consists of the following:

a. **Briquette mould:** The mould is made of brass metal with shape and dimensions as shown in Fig. 20.1. Its two ends are called clips. It consists circular holes to grip the fixed and movable ends of the testing machine. The mould when properly assembled form a briquette specimen of following dimensions:

$$\text{Length} = 75 \text{ mm}$$
$$\text{Distance between clips} = 30 \text{ mm}$$
$$\text{Width at month of clips} = 20 \text{ mm}$$
$$\text{Cross-section at minimum width} = 10 \quad 10 \text{ mm}$$
$$\text{Thickness throughout} = 10 \text{ mm}$$

b. **Water bath:** A bath preferably with thermostatically control to maintain the specified test temperature. It contains not less than 10 litres of water. It allows a depth of immersion of specimen not <100 mm and supported on a perforated shelf not <50 mm from the bottom of bath.

c. **Testing machine:** It is used for pulling the briquette of bituminous material apart. Specimen will be continuously submerged in water while the two clips are being pulled apart.

d. **Thermometer:** It should have range 0–44 C (Figs 20.1 and 20.2).

20.5 PROCEDURE

Refer IS 1208 : 1978

1. Melt the bitumen sample at temperature of 75–100 C above the approximate softening point until it is fluid.
2. Strain the fluid through IS sieve 30. After that stirr the fluid and then pour it in the mould and place it on a plate.
3. After 30 to 40 minutes, keep the plate assembly along with the sample in a water bath, maintained at 27 C for 30 minutes.
4. Remove the sample and mould assembly from water bath and then cut off the excess bitumen by levelling the surface using hot knife.

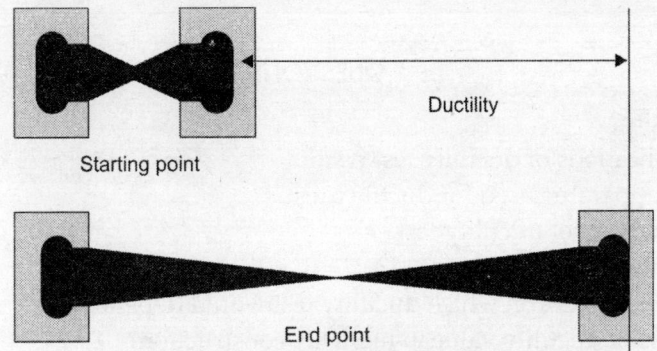

Fig. 20.1: Concept of ductility test

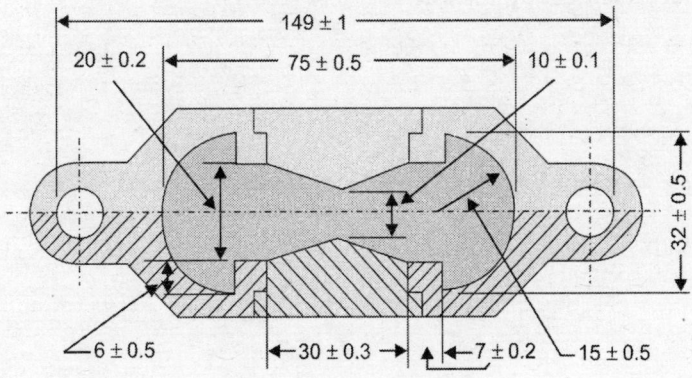

Fig. 20.2: Briquette mould

5. After trimming the specimen, replace the sample and mould in water bath maintained at 27° for 85 to 95 minutes.
6. Remove the sides of the mould and than hook the clips carefully on the machine without causing initial strain.
7. Set the pointer to read zero. Start the machine and then pull the two clips horizontally.
8. Record the distance (in cm) at which the bitumen thread of each specimen breaks.
9. From the observations recorded, compute the ductility of the binder (bitumen). The mean of two observations, rounded to nearest whole number gives the ductility value (Table 20.1).

20.6 OBSERVATIONS AND CALCULATIONS

Table 20.1: Ductility test

Test property	Briquette No.		
	1	2	3
Initial reading, Di			
Final reading, Df			
Ductility value = Df-Di (cm)			
Mean ductility value, cm			

QUESTIONS

1. What is ductility?
2. Write the applications of ductility test results.
3. What are the apparatus used for ductility test?
4. Write the procedure of ductility test.
5. How is the ductility value expressed?
6. Briefly write the factors on which ductility test would depend?
7. What is the use of ductility value in highway construction?
8. Explain ductility of bitumen and its significance.
9. How is ductility value expressed?
10. What are the factors affecting the ductility test results?

CHAPTER 21

Penetration Test

21.1 OBJECT

To determine the penetration in bitumen, i.e. to determine consistency of bituminous material.

21.2 THEORY

The consistency of bitumen is determined by penetration test which is a very simple test. The basic principle of the penetration test is the measurement of the penetration of a standard needle in a bitumen sample maintained at 25 C during 5 seconds, i.e. the penetration of a bituminous material is the distance in tenths of a millimetre that a standard needle will penetrate vertically into a sample of the material under standard conditions of temperature, load and time.

$$\text{Penetration} = \text{Distance drop (in mm)} \quad 10$$

21.3 APPLICATIONS

Penetration test grades the bitumen material in terms of its hardness. A 50/100 bitumen denotes that penetration value ranges between 50 and 100. For bituminous macadam and penetration macadam, Indian Roads Congress suggests bitumen grades 30/40, 60/70 and 80/100. If the regions are warmer lower grades are preferred and if it is colder bitumen with higher grades are preferred.

21.4 APPARATUS

According to ISI, the following are the apparatus:
  i. **Container:** A flat bottomed cylindrical metallic container of diameter 55 mm and height 35 mm or 57 mm.
 ii. **Needle:** A straight, highly polished cylindrical hard steel needle with conical end.
iii. **Water bath:** A water bath maintained at 25 ± 1 C containing not less than 10 litres of water, the sample is immersed to depth not less than 100 mm from the top and supported on a perforated shelf not less than 50 mm from the bottom of the bath.
 iv. **Transfer dish or tray:** A small tray which can keep the container fully immersed in water during the test:
  v. **Penetrometer:** The penetrometer should be such that it will allow the needle to penetrate without much friction and the dial is accurately calibrated to give penetration value in units of one-tenth of 1 mm.

vi. **Thermometer** of range 0–44 C.

vii. Time measuring device: With an accuracy ± 0.1 second (Figs 21.1 and 21.2).

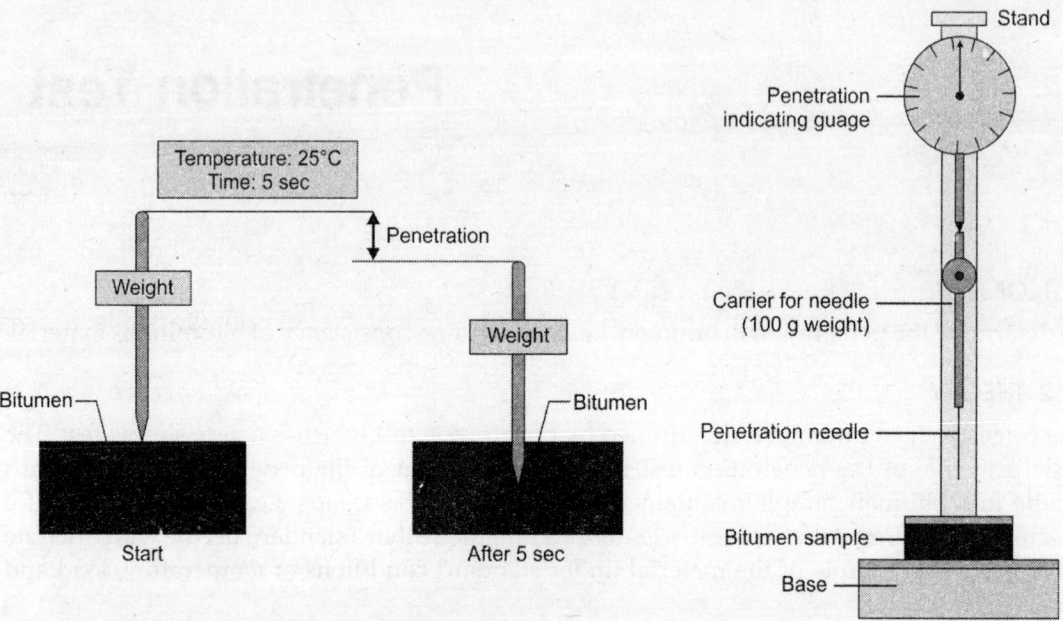

Fig. 21.1: Penetration of needle in bitumen   Fig. 21.2: Penetration test apparatus

21.5 PROCEDURE

Refer IS 1203: 1798

1. Soften the bitumen to a pouring consistency between 75 and 100 C above the approximate temperature at which bitumen softens.
2. Stir the sample material thoroughly to make it homogenous and free from air bubbles and water.
3. Pour the sample material into the container to a depth at least 15 mm more than the expected penetration.
4. Cool the container in atmosphere of temperature not less than 13 C for one hour. Then place the container along with the transfer dish in the water bath to 25 ± 1 C for 1–1.5 hour.
5. Bring the needle of penetrometer in contact with the surface of the test specimen.
6. The initial reading of the penetrometer dial is either adjusted to zero or the initial reading noted.
7. Release the needle and immediately start the stopwatch. Release the needle exactly for 5.0 seconds. Not the final reading of penetrometer dial gauge.
8. Make at least 3 readings at points on the surface of the sample not less than 10 mm apart and not less than 10 mm from side of dish.
9. Find difference between the initial and final readings. It gives penetration value.
10. After the first test clean the needle with benzene and dry it.

11. Repeat the test second time with fresh sample and find the penetration value.
12. Find the mean value of penetration for each of the test separately.

21.6 PRECAUTIONS

1. Clean the needle before performing the test.
2. Put the penetration needle not <10 mm from side of the container.
3. Avoid any movement of container during testing.
4. Keep the sample free from any extraneous matter.

21.7 OBSERVATIONS AND CALCULATIONS

Table 21.1 shows penetration test.

Table 21.1: Penetration Test

Test property	Briquette No.		
	1	2	3
Initial reading, Di			
Final reading, Df			
Ductility value = Df-Di (cm)			
Mean ductility value, cm			

QUESTIONS

1. Consistency of bitumen is determined with what test?
2. What is the basic principle of penetration test?
3. Write the expression for penetration.
4. Write the application of penetration test.
5. Write the apparatus used for penetration test.
6. Write the procedure of penetration test.
7. What precautions will you take while performing penetration test?
8. What is meant by 80/100 grade bitumen?
9. What does the penetration value of bitumen indicate?
10. What are the possible effects of penetration test conducted at higher temperature?
11. What parameter does penetration test identify?
12. How is penetration value of bitumen expressed?
13. What are the standard load, time and temperature specified for penetration test?

CHAPTER 22

Specific Gravity Test for Bitumen

22.1 OBJECT
To determine specific gravity of bitumen.

22.2 THEORY
The specific gravity is defined as ratio of mass of a given volume of bitumen material to the mass of an equal volume of water, the temperature of both being specific as 27 C ± 0.1 C. Bituminous binders are most commonly characterized by their physical properties. The physical properties directly describe the performance as a constituent in conventional flexible pavement construction. The specific gravity gives an idea about the actual physical properties of the material.

22.3 APPLICATIONS
The Indian Standard Institution specifies that the minimum specific gravity values of paving bitumen at 27 C shall be 0.99 for grades A25, A35, A45, A65, S35, S45 and S65, 0.98 for A90 and S90 and 0.97 for A200 and S200.

22.4 APPARATUS
1. **Specific gravity bottle**: Two different types of bottles of 50 ml capacity for each.
2. **Constant temperature bath:** The bath must be large enough to hold pycnometer bottle. The bath should be able to maintain a test temperature.
3. **Thermometer:** Capable to measure 0–44 C.
4. **Funnel:** Funnel with diameter of neck smaller than the diameter of specific gravity bottle should be taken.
5. **Weighing balance:** A balance of capacity of 400 g, accurate to 0.001%.
6. **Cloth:** Dry soft absorbent cloth.

22.5 PROCEDURE
1. Clean, dry and weigh the specific gravity bottle alongwith the stopper. Let it be W_1.
2. Completely fill the pycnometer bottle with distilled water along with stopper. Wipe moisture from surface with clean , dry cloth. Weigh the pycnometer filled with water with stopper. Let it be W_2.

3. Empty the bottle and than dry it. Heat the bituminous material to pouring temperature. Pour it in the empty bottle to fill approximately half of its volume. Keep the partly filled bottle in water bath at 27 C temperature for half an hour.

4. Take out the pycnometer bottle out of water bath and then wipe moisture from the surface of bottle. Take the weight of pycnometer bottle along with the material and the stopper. Let it be W_3.

5. Fill the empty portion of bottle with distilled water along with stopper. There should not be any air bubbles in the bottle. Wipe all the surface water with clean, dry cloth. Take the weight of the bottle along with material, water filled and stopper. Let it be W_4.

22.6 OBSERVATIONS AND CALCULATIONS

Table 22.1 shows specific gravity test for bitumen.

$$\text{Specific gravity} = \frac{(W_3 - W_1)}{(W_2 - W_1) - (W_4 - W_3)}$$

Table 22.1: Specific gravity test for bitumen

Bitumen grade =   Test temperature

Observation No.		Sample No. 1		Average value
	1	2	3	
Weight of bottle = W_1 g				
Weight of bottle + distilled water = W_2 g				
Weight of bottle + half filled material = W_3 g				
Weight of bottle + half filled material + distilled water = W_4 g				
Specific gravity = $\dfrac{(W_3 - W_4)}{((W_2 - W_1) - (W_4 - W_3))}$				

22.7 PRECAUTIONS

1. Make the surface of sample free from sticking of air bubbles.
2. Maintain the specified temperature during weighing.
3. Use only cooled distilled water.
4. Take the weight carefully accurate to 0.001 g.
5. Make at least there measurements for determining the value of specific gravity.

QUESTIONS

1. Define specific gravity.
2. What are the apparatus for specific gravity test?
3. What are the applications of specific gravity?
4. Write procedure of specific gravity test of bitumen.
5. What is the importance of specific gravity test of bitumen?

CHAPTER 23

Viscosity Test

23.1 OBJECT

To determine the viscosity of bitumen binder.

23.2 THEORY

Viscosity of a fluid bitumen is the property due to which it offers resistance to flow. Lower the viscosity, higher will be movement of fluid and higher the viscosity, the slower will be movement of fluid. The binder spreads, moves into and fills up the voids between aggregates. Highly viscous binder cannot fill up the voids completely and when viscosity is lower the binder cannot hold the aggregates together. The viscosity is defined as the time taken in seconds by 50 cc of material to flow from a cup through a specified orifice under standard conditions of test and at specified temperature.

23.3 APPARATUS

As per IS 1206 (Part I): 1978 following equipment are required:

Tar cup: The tar cup is made of hard brass with orifice diameter 10 mm or 4 mm ± 0.025 mm. It is fitted with an external brass collar at the upper end.

Valve: It serves to close the orifice of the cup. It is a metal rod made of phosphore-bronze having a sphere at the bottom end to close the orifice and a levelling peg at the upper end for support.

Water bath: It is made of copper sheet, is cylindrical in shape, about 160 mm in diameter and 105 mm in depth. It is mounted on three equidistant legs.

Sleeve: To receive the cup and to hold it in position.

Stirrer: It consists of four vertical vanes.

Curved shield: It is fixed to the upper edge of the cylinder and extends to within about 5 mm of the walls of the water bath. This shield carries an insulated handle for rotating the stirrer, a support for a thermometer, and a support for the valve.

Receiver: A 100 ml graduated measuring cylinder having markings on 25 and 75 ml levels.

Thermometers: Two standard thermometers having graduations 10–100 C. One for water bath and another for cup.

Stopwatch: Stopwatch with least-count of 0.5 second (Fig. 23.1).

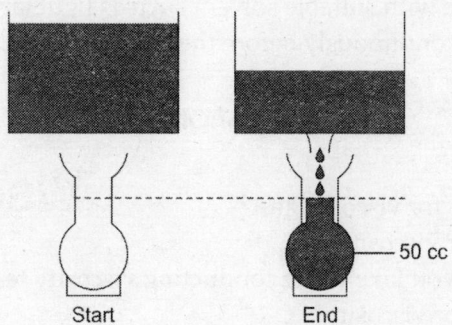

Fig. 23.1: Viscosity test concepts

23.4 PROCEDURE

Refer IS 1206 (Part III): 1978

1. Level the tar cup property by means of the levelling screws provided at the bottom of legs.
2. Heat the water in the water wath to the temperature specified for the test and maintain throughout the test by stirring the water continuously.
3. Clean the tar cup and orifice with a suitable solvent and dry thoroughly.
4. Heat the sample material at temperature 20°C above the temperature of test and allow the material to cool. During this stirr the material continuously.
5. When the temperature is cooled to a temperature which is slightly above test temperature, pour the material into the cup until the levelling peg on the valve rod is just immersed.
6. Fill the measuring cylinder up to 20 cc with mineral oil or 1% by weight of soap solution. Place the measuring cylinder (receiver) under the orifice.
7. Place the thermometer in the cup and stirr the material in the cup until the temperature is within ± 0.2 of the temperature specified for the test.
8. Open the value when the sample material reaches the specified testing temperature within ± 0.1°C and is maintained for 5 minutes.
9. Start the stopwatch when the measuring cylinder records 25 ml (cc), i.e. then stop the watch when the measuring cylinder records 75 ml. Record the time in seconds.
10. Report the viscosity as the time taken in seconds by 50 ml of material to flow out at the temperature specified for the test (Table 23.1).

23.5 OBSERVATIONS AND CALCULATIONS

Table 23.1 Viscosity test

Test temperature	Test I	Test II
Time taken to flow 50 cc of the binder		
Viscosity		

Mean viscosity = Seconds

23.6 PRECAUTIONS

1. Test the orifice size at frequent intervals with a gauge having appropriate diameters.
2. Level the tar cup properly.

3. Clean the tar cup orifice with suitable solvent such as light tar oils and dry thoroughly.
4. Stir the test specimen continuously before the raising of value.

QUESTIONS

1. What is viscosity?
2. What are the apparatus for viscosity test?
3. Write the procedure for viscosity test.
4. What precautions will you take while conducting viscosity test?
5. What is absolute unit for viscosity?
6. What is the viscosity of bitumen at softening point?
7. Do viscosity and penetration tests actually signify same property of bitumen?

CHAPTER 24

Flash Point and Fire Point Test

24.1 OBJECT

To determine flash point and fire point of bituminous material.

24.2 THEORY

According to IS 1209: 1958 the flash point of a material is the lowest temperature at which the vapour of substance momentarily catches fire when it comes in contact of flame, in the form of a flash under specified conditions of test. The fire point is the lowest temperature at which the material gets ignited due to flame and burns at least for 5 seconds under specified condition of test.

24.3 APPLICATIONS

Different types of bituminous material have varying flash and fire points. Utmost care should be taken while heating bitumen before mixing or application. The bitumen should be heated below the flash point. The minimum value of flash point as per ISI is 175 C for all grades of bitumens.

24.4 APPARATUS

Refer IS 1209 : 1978

Pensky Martens Closed Tester having: (i) cup, (ii) lead, (iii) stirring device, (iv) cover, (v) shutter, (vi) flame exposure device, etc.

24.5 PROCEDURE

Refer IS 1209 : 1978

1. Clean the cup and dry it thoroughly before the start of the test.
2. Fill the bitumen material up to filling mark.
3. Insert thermometer into the material so that the bulb of thermometer penetrates at least 45 mm.
4. Heat the bitumen sample. Stir the bitumen in the cup at approximately 60 revolutions per minute. The heating is done at the rate of 5 to 6 C per minute.
5. Note the temperature at which first flash appears when test flame is brought close to the surface of material. This temperature is noted as flash point temperature.

6. After flash point is obtained, heating should be continued at such a rate that the increase in temperature recorded by the thermometer is neither <5 C nor >6 C per minute.
7. Light a test flame and adjust it so that it is of size of a bead 4 mm in diameter.
8. Note the temperature at which the application of test flame causes the material to ignite and bum for at least 5 seconds. This temperature is noted as fire point temperature.

24.6 OBSERVATIONS AND CALCULATIONS

Table 24.1 shows flash point and fire point.

Table 24.1 Flash point and fire point

Test	Test number			Mean
	1	2	3	
Flash point				
Fire point				

QUESTIONS

1. What is flash point and fire point of bitumen material?
2. Define flash point and fire point of bitumen.
3. What are applications of flash point test?
4. What apparatus are used for flash point and fire point tests?
5. Write procedure of flash point and fire point test.

CHAPTER 25

Solubility of Bituminous Material in Carbon Disulphide or Trichloroethylene

25.1 OBJECT

To determine bitumen content of a bituminous material by means of its solubility in carbon disulphide or trichloroethylene.

25.2 THEORY

Pure bitumen is used as binding material for flexible pavement. Impurities in bitumen are harmful constituent of the bituminous mixture and are hazardous for flexible pavement. The portion of bitumen which is soluble in carbon disulphide gives the pure bitumen content.

25.3 APPARATUS

Refer IS 1216 : 1978
- Gooch crucible
- Conical glass flask having 200 ml capacity
- Oven maintaining maximum temperature of 150°C
- Suction pump
- Electronic weighing balance accurate to 0.001 g
- Carbon disulphide conforming to IS 717: 1969
- Trichloroethylene conforming to IS 245: 1970.

25.4 PROCEDURE

Refer IS 1216 : 1978
1. About 2 g of dry bitumen is weighed into a 200 ml conical flask.
2. 100 g of carbon disulphide or trichloroethylene is then added into the conical flask.
3. The contents of the flask is then stirred and allowed to stand for a period of one hour at room temperature.
4. The contents of the flask is filtered through Gooch crucible maintaining the filtration rate of two drops per second. All insoluble material present in the flask is also transferred to the Gooch crucible.
5. The material retained in the crucible is washed with successive small amount of carbon disulphide or trichloroethylene. This process is continued up to time when the liquid obtained after filtration is not discolored (Table 25.1).

6. The crucible is dried in air for 30 minutes and then its is placed in oven which is at temperature 100–110 C for 60 minutes.
7. The crucible is cooled in the desiccators.
8. After cooling in desiccators the crucible is weighed.

25.5 OBSERVATIONS AND CALCULATIONS

The bitumen soluble in carbon disulphide or trichloroethylene

$$= \frac{W_1 - W_2}{W_1} \times 100\%$$

where W_1 = Weight in g of dry bitumen taken for test.

           W_2 = Weight in g of insoluble material retained in the Gooch crucible

Table 25.1: Solubility test

Observations No.	Observations number	Mean
Weight in g of dry bitumen taken for test W_1 (g)		
Weight in g of insoluble material retained in the Gooch crucible, W_2 (g)		
Bitumen (pure) = $\dfrac{W_1 - W_2}{W_1} \times 100\%$		

25.6 PRECAUTIONS

1. Carbon disulphide or trichloroethylene should be kept away from flame or any other heat source.
2. Adequate ventilation is provided to save persons from vapours which is toxic.
3. The sample should be free from water.

QUESTIONS

1. What is pure bitumen?
2. What apparatus are used for solubility test?
3. Write the procedure of solubility test.
4. What precautions should be taken while performing solubility test?
5. Write expression for pure soluble bitumen.

CHAPTER 26

Loss in Heating

26.1 OBJECT

To determine the loss in weight of bitumen due to heating.

26.2 THEORY

Oxidation, evaporation and cold temperature cause hardening of bitumen. The properties like penetration, softening point, ductility are lost when bitumen becomes hard. Loss of heating is a simple test to determine the loss in weight of bitumen.

26.3 APPARATUS

Refer IS 1212 : 1978

- *Oven:* It has a double-walled electrically heated convection type chamber of rectangular shape. Its height is minimum of 292 mm, width and depth each not <298 mm.
- *Perforated metal shelf:* A perforated aluminium shelf of approximately 250 mm diameter placed in centre of oven with mechanical means for rotation at the rate of 5 to 6 revolutions per minute.
- *Thermometer:* Standard thermometer which can measure temperature up to 163 C.
- *Container:* Heat resistance glass cylinder with flat bottom having internal diameter 55 mm and internal depth 35 mm.

26.4 PROCEDURE

1. Refer IS 1212: 1978
2. It should be observed that the sample is free of water. The sample must be heated carefully at 135 ± 5.5 C in a covered container.
3. The container is heated in an oven at 100–110 C for 30 minutes and cooled at room temperature. It is then weighed.
4. When the sample becomes as fluid, 50 ± 0.5 g of sample is poured into each of the specified container for tests.
5. The oven is heated. When temperature of oven becomes 163 ± 1 C, the container is placed on the revolving shelf. The shelf is rotated during entire test at a rate of 5 to 6 rotations per minute. The temperature of oven is maintained at 163 ± 1 C for five hours.
6. When the heating period is over, the containers are removed. They are then cooled at room temperature. They are then weighed.

26.5 OBSERVATIONS AND CALCULATIONS

Initial weight of bitumen = W_1 (g)

Final weight of bitumen after heating = W_2 (g)

$$\text{Loss of heating} = \frac{W_1 - W_2}{W_1} \times 100\%$$

Table 26.1 shows loss in weight after heating.

Table 26.1: Loss in weight after heating

Observation	Observation number			Mean
	1	2	3	
Initial weight of bitumen, W_1 (g)				
Final weight of bitumen after heating, W_2 (g)				
Loss in weight = $\dfrac{W_1 - W_2}{W_1} \times 100\%$				
Mean value				

26.6 PRECAUTIONS

1. The loss on heating test should be conducted in duplicate.
2. Specified time and temperature should be considered carefully.

QUESTIONS

1. What is objective of the heating test?
2. Write expression for loss on heating.
3. What are the apparatus used for test 'loss on heating'?
4. Write procedure of the test 'loss on heating'.
5. Write precautions to be taken for 'loss on heating' test.

CHAPTER 27

Fraass Breaking Point Test

27.1 OBJECT

To measure the temperature at which bitumen reaches a critical stiffness and cracks.

27.2 THEORY

In India, there is significant variation in daily and seasonal temperature of the pavement. When temperature is low flexibility, stiffness and cyclic properties of pavement material (mixes) change significantly.

This change causes fatigue cracking, premature cracking, thermal cracking and rutting. Hence, it is necessary to determine the behaviour of bitumen at low temperature.

27.3 APPARATUS

- Bending apparatus
- Plaque
- Cooling arrangement
- Thermometer.

27.4 PROCEDURE

Refer IS 9381: 1978

1. The annular space is filled with acetone up to about half of its height.
2. The plaque is fixed between the clips of the bending apparatus.
3. The solid carbon dioxide is mixed with acetone through the funnel at such a rate that the temperature falls at the rate of 1 C per minute.
4. The plaque is bended once every minute by turning the handle at a rate of one revolution/ second and until it is checked. Then it is turned backwards at the same speed.
5. The temperature at which one or more cracks appear on sample is noted. This temperature gives the breaking point.

27.5 PRECAUTIONS

1. The plaque should be uniformly coated by bitumen sample.
2. The rate of cooling should be uniform.
3. The temperature should be noted properly at breaking point, i.e. when one or more cracks appear on the sample.

27.6 OBSERVATIONS AND CALCULATIONS

Table 27.1 shows fraass breaking point test.

Table 27.1 Fraass breaking point test

Observation No.	Observation number			Mean
	1	2	3	
Temperature at which cracks appear [C]				

Result

Fraass breaking point = C

1. Define Fraass breaking point.
2. Write apparatus used for Fraass breaking point test.
3. Write procedure for Fraass breaking point test.
4. What precautions will you take while conducting Fraass breaking point test?
5. The temperature at which crack develops on sample is termed what?

CHAPTER 28

Separation Test for Modified Bitumen

28.1 OBJECT

To evaluate the separation of modifier and bitumen from modified bitumen during hot storage by comparing the ring and ball test.

28.2 APPARATUS

- *Aluminium tubes:* 25.4 mm diameter and approximately 540 mm height.
- *Oven:* Capable of maintaining 163 ± 5 C temperature.
- *Freezer:* Capable of maintaining 6.7 ± 5 C temperature.
- *Rack type arrangement:* Capable to support the aluminium tube.
- *Cutter:* Capable of cutting the aluminium tube.

28.3 PROCEDURE

1. Approximately 50 g of material should be heated carefully around 135 ± 5.5 C until material becomes fluid.
2. One end of the empty tube is sealed and then the tube is placed on the rack with its sealed end.
3. The prepared sample is poured into the tube. It is then placed with rack in oven at 163 ± 5 C.
4. The tube with rack is kept in oven for 48 ± 4 hours. After that the rack is removed with tube from oven and placed in the freezer at 6.7 ± 5 C. The tube must be in vertical position. The tube must be in freezer for minimum of 4 hours unit the sample becomes solid.
5. The tube is then removed from freezer and kept on flat surface. The tube is then cut into three equal parts.
6. The top and bottom portion of tube are put in separate beakers.
7. The beakers are then placed into an oven at 163 ± 5 C up to such a duration that bitumen becomes fluid. The samples are stirred thoroughly. Then the top and bottom samples are placed into respective rings and then softening point test is conducted. The top and bottom samples from the same tube are tested at one time.
8. The difference of temperature in C of softening points of top sample and bottom sample gives the separation difference.

28.4 PRECAUTIONS

1. Only end samples should be taken for softening point test.
2. The tube sample when filled should be free from air pocket.
3. The test temperature must be uniform.
4. When softening point test is conducted, the rate of heating must be uniform.

28.5 OBSERVATIONS AND CALCULATIONS

Table 28.1 shows separation test.

Table 28.1: Separation test

Date:				Type of bitumen/bitumen grade
Softening point value, C		*Observation number*		*Mean C*
	1	*2*	*3*	
Sample collected from top to tube C, T_1				
Sample collected from bottom to the tube C, T_2				
Difference between the top and bottom sample				
$T = (T_1 - T_2)$ C				

QUESTIONS

1. What is the objective of separation test?
2. What are the apparatus used for separation test?
3. Write the procedure of separation test.
4. What are the precautions to be taken while conducting separation?
5. What does separation actually mean?

CHAPTER 29

Elastic Recovery Test

29.1 SCOPE

This test is commonly used in quality control for modified bitumen in flexible pavement construction to find out the elastomeric characteristic of the additive used to modify the conventional bitumen. This is intended to assess the degree for bitumen modification by elastomeric additive.

29.2 APPARATUS

- *Testing machine:* For pulling the briquette of material apart.
- *Mould:* Made of brass.
- *Hot oven:* It should be capable to maintain 135 ± 5.5 C temperature.
- *Thermometer:* Standard thermometer to measure the temperature from –8 to 32 C.
- *Scissors:* Scissors must be capable of cutting modified bitumen at test temperature.
- *Scale:* Transparent scale capable to measure ductility up to 250 mm with ± 1 mm accuracy.

29.3 PROCEDURE

1. The sample is prepared and conditioned as per the procedure of the ductility test.
2. The test specimen is elongated at specified rate of 50 ± 2.5 mm per minute at specified temperature (15 C) to a deformation of 10 cm.
3. As soon as the specimen is elongated to a deformation of 10 cm, the specimen is cut into two halves at mid-point using scissors.
4. The specimen is kept in water bath in an undisturbed condition for a period of one hour at specified temperature.
5. After one hour, the travelling carriage is moved carefully back into the position near the fixed half of the specimen so that the two pieces of modified bitumen just touch each other.
6. The length of recombined specimen is measured as 'y' cm. Let the original elongation of specimen be x.

29.4 OBSERVATIONS AND CALCULATIONS

$$\text{Elastic recovery} = \frac{x - y}{10} \times 100\% = \frac{10 - y}{10} \times 100\%$$

29.5 PRECAUTIONS

1. The temperature should be maintained properly.
2. The test specimen should be submerged in water while conducting the test.
3. The machine should not vibrate at the time of pulling operation.
4. The two pieces of modified bitumen should just touch each other and then reading should be taken.

QUESTIONS

1. Write the significance of recovery test.
2. What apparatus are used for recovery test?
3. Write the procedure for conducting recovery test.
4. What precautions will you take while conducting recovery test?

CHAPTER 30

Marshall Stability Test

30.1 OBJECT

To determine Marshall's stability value and flow value for the given bituminous mixture.

30.2 THEORY

The Marshall stability of the mix is defined as a maximum load carried by a compacted specimen at a standard test temperature at 60°C. The flow value is the deformation the Marshall test specimen undergoes during the loading up to maximum load, in 0.25 mm units (Fig. 30.1).

30.3 APPARATUS

- *Mould assembly:* It consists of cylindrical moulds of diameter 10 cm and height 7.5 cm. It further consists base plate and extension collars. They are designed to be interchanged with either end of cylindrical moulds.
- *Sample extractor:* It is used to extract the compacted specimen from the mould.
- *Compaction hammer:* It has a flat circular tamping face, weight 4.5 kg with free fall of 45 cm.
- *Compacting pedestal:* It consists of a 20 × 20 × 45 cm wooden block capped with a 30 × 30 × 2.5 cm MS plate to hold the mould assembly in position during compaction.
- *Breaking head:* It consists of upper and lower cylindrical segments or test heads having an inside radius of curvature 5 cm. The lower segment is mounted on a base having two perpendicular guide rods which facilitate insertion in the holes of the upper test segment.
- *Loading machine:* The loading machine is provided with a gear system to lift the base in upward direction. Precalibrated proving ring of 5 tonne capacity is fixed on the upper end of the machine. The specimen contained in the test head is placed in between base and the proving ring. The loading jack produces a movement at the rate of 5 cm per minute. Machine is capable of reversing its movement downward also. This facilitates adequate space for placing test head system after one specimen has been tested.
- *Flow meter:* The flow meter consists of a guide sleeve and a gauge. The activating pin of gauge slides inside the guide sleeve. Least count of 0.25 mm is adequate. The flow value refers to the total vertical upward movement from initial position at zero load to a value at maximum load. The dial gauge or flow meter should be able to measure accurately the total vertical movement upward.

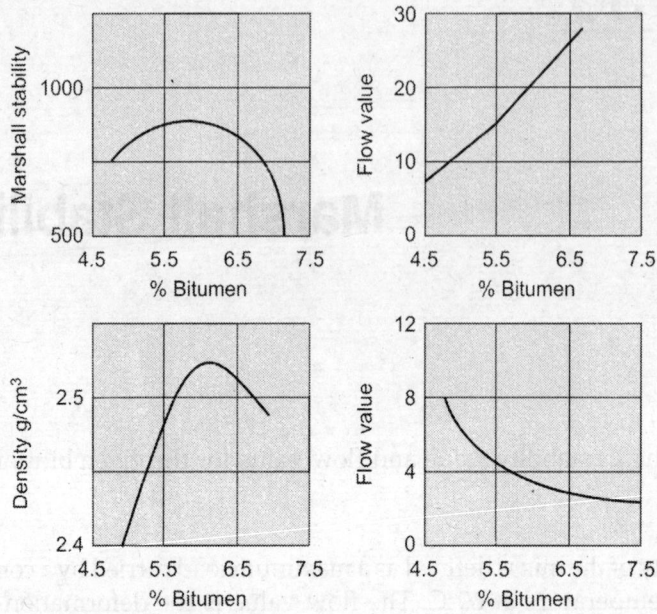

Fig. 30.1: Bituminous mix design by Marshall test

In addition to above apparatus following apparatus are also required:

- Oven or hot plates
- Mixing apparatus
- Water bath
- Thermometers of range up to 200 C with an accuracy to 2.5 C.

30.4 PREPARATION OF TEST SPECIMEN

1. Take approximately 1200 g of aggregate and filler. Heat the aggregate in the oven to the mixing temperature.
2. Add bitumen to aggregate at mixing temperature for 80/100 grade bitumen may be around 154 C and that for 60/70 grade about 160 C. Mix the materials in a heated pan with heated mixing tools.
3. Return the mixture to oven and reheat it to the compacting temperature. The compacting temperatures may be about 138 C for 80/100 grade bitumen and 149 C for 60/70 grade.
4. Place the mix in a heated Marshall mould with collar and base. Compact the mix by rammer with 50 blows of hammer on either side.
5. When the compaction is over, invert the mould such that collar is on bottom and base plate at the top. Remove the base plate and then extract the sample by extraction. Allow the sample to stand for a few hours to cool.
6. Obtain the sample's mass in air and submerged to measure density of specimen.

30.5 PROCEDURE

1. Heat the specimens either in water bath for 30–40 minutes or in an oven for minimum of two hours at temperature 60 ± 1 C.

2. After removing the specimen from water bath or oven, place the specimen on lower segment of the breaking head. Then place the upper-segment of breaking head on the specimen. Keep the complete assembly in position on the testing machine.
3. Adjust the deformation measuring gauge, i.e. the flow meter to zero reading.
4. Get the maximum load by applying load at a rate of 50 mm per minute and get the reading of dial gauge, i.e. the flow meter in units of mm. Measure the maximum load in Newtons (N).

30.6 OBSERVATIONS AND CALCULATIONS

Table 30.1 shows Marshall stability value.

$$\text{Let bitumen content} = W_b\%$$
$$\text{Aggregate} = W_A\%$$
$$\text{Specific gravity of bitumen} = G_b$$
$$\text{Specific gravity of aggregates} = G_a$$
$$\text{Density of compacted mix} = y_d \text{ g/cm}^3$$

$$\text{Specific gravity of mix} = \frac{100}{\dfrac{W_b}{G_b} + \dfrac{W_A}{G_A}}$$

$$\text{Volume of bitumen VB} = \frac{W_b \times y_d}{G_A}\%$$

$$\text{Volume of aggregates, VA} = \frac{(100 - W_b) \times y_d}{G_A}$$

$$\text{Voids in mineral aggregates VMA} = 100 - \text{VA}$$

Table 30.1: Marshall stability value

Observations	Sample 1	Sample II	Sample III
Density of compacted mix, r_d, g/cm^3			
Specific gravity ox G_M =			
Volume of bitumen, V_B = %			
Volume of aggregates, V_A = %			
Voids in mineral aggregates, VMA = $(100 - V_A)$			
Voids in mix = V_M = $100 - (V_B + \text{VA})$			
Voids filled with bitumen = VFB =			
Measured stability = N			
Flow value = mm			

$$\text{Voids in mix} = V_M$$
$$= 100 - (V_A - V_B)$$

$$\text{Voids filled with bitumen} = \text{VFB} = \frac{100 \times V_B}{VMA}$$

$$\text{Measured Stability} = N$$
$$\text{Flow value} = \text{mm}$$

QUESTIONS

1. What is the object of Marshall stability test?
2. What are the apparatus used for Marshall stability test?
3. How will you prepare sample for Marshall stability test?
4. Write in brief the procedure for Marshall stability test.
5. Write expression for specific gravity of mix.
6. Write expression for volume of bitumen.
7. Write expression for volume of aggregate.
8. Write expression of voids in mineral aggregates (VMA).

PART 5

Field Tests

CHAPTER 31

Test on Bitumen Emulsion and Cutback Bitumen

31.1 OBJECT

To perform tests on bitumen emulsion and cutback bitumen.

31.2 THEORY OF BITUMEN EMULSION

A bitumen emulsion is a liquid product in which a substantial amount of bitumen is suspended in a finely divided condition in anaqueous medium. It is stabilized by means of one or more suitable stabilizers. The desired type of emulsion depends on the aggregate, climatic conditions, environmental conditions, etc. The emulsion are prepared by: (i) Colloid mill method and (ii) The high speed mixture method. In air tight drums, the manufactured emulsion is stored. The advantages of bitumen emulsions are: (i) they can be used without heating of spraying or preparing mixes, (ii) they are useful for patch repair works.

Types of Bitumen Emulsions

The bitumen emulsion are of five types as follows:
  i. Rapid setting type, RS-1 and RS-2
 ii. Medium setting type, MS
iii. Slow setting types, SS-1 and SS-2.

31.3 APPLICATIONS

1. *Rapid setting bitumen emulsions:* They are used in spray applications like tack coat, for surface treatments, surface dressing and penetration macadam.
2. *The medium setting emulsion:* The medium setting emulsion be used in cold bitumen mixes in which the percentage of coarse aggregates is present.
3. *The slow setting emulsions:* They are used for prime coat, slurry seal treatments, recycling works and in soil stabilization. They are also used with well graded bituminous mixes containing a substantial proportion of fine aggregates.

31.4 TEST ON BITUMEN EMULSION

The following are the test for bitumen emulsions:
  i. *Viscosity test:* It is conducted to assess ability to be spread through test.
 ii. *Watercontent:* It is performed to estimate the actual binder quantity.
iii. *Settlement test:* It is performed to evaluate settlement when left standing.

iv. *Demulsibility test:* It is performed to find the residue after mixing with calcium chloride as specified.

v. *Miscibility in water:* It is performed to assess coagulation due to addition of distilled water.

vi. *Cement mixing test:* It is performed to assess stability in presence of fine aggregates.

vii. *Coating test:* It is performed to assess coating of stone aggregates.

viii. *Sieving test:* It is performed to measure sedimentation of emulsion during storage.

ix. *Particle charge:* It is performed to evaluate the type of charge.

31.5 PROPERTIES AND REQUIREMENTS

The physical and chemical requirements of the three types of emulsions, RS, MS, SS are given in Table, Appendix IV.

31.6 CUTBACK BITUMEN

Cutback bitumen is obtained by blending bitumen binder with suitable volatile diluents. After the cutback mix is used in construction work, the volatile solvent gets evaporated, the binder starts hardening. The rate at which the cutback hardens, The rate of hardening depends on the volatile oil used as diluents and atmospheric temperature.

31.6.1 Types of Cutback Bitumen

The cutback bitumen are of three types follows:

i. Rapid curing (RC)

ii. Medium curing (MC)

iii. Slow curing (SC).

The rapid curing cutback bitumen are classified by BIS, on the basis of initial kinematic viscosity into a four grades with designations RC-70, RC-250, RC-800 and RC-3000 in the increasing order of initial viscosity. RC-70 is rapid curing product of sufficiently low initial viscosity to be sprayed at normal air temperature without heating, whereas the RC-800 and RC-3000 are products of high viscosity which cannot be easily mixed with fine aggregate of soil at low temperature. Medium curing (MC) cutback bitumen are classified on the basis of initial viscosity into fine grades MC-30, MC-70, MC-250, MC-800 and MC-3000 in the increasing order of viscosity. MC-30 is to be used as primer. Similarly, the slow curing (SC) cutbacks are classified into four grades with designation SC-70, SC-250, SC-800 and SC-3000.

31.6.2 Tests

The following tests are performed on cutback bitumen.

1. Kinematic viscosity

2. Flash point test

3. Distillation test to find fractions of distillage up to 190, 225, 260, 315 and 360 C

4. Tests on residue from distillation up to 360 C

5. Viscosity at 60 C

6. Ductility at 27 C

7. Matter soluble in trichloro-ethylene

8. Water content.

31.6.3 Properties

RC, MC and SC types of cutback bitumen of various grades should comply with the requirements with regard to the properties such as viscosity at different test temperatures, flash point, distillation fractions, residue from distillation up to the specified temperature and tests on residue from distillation. The requirements of the various grades of cutback bitumen as specified by BIS are given in Tables in Appendix IV.

QUESTIONS

1. What is a bitumen emulsion?
2. What is a object of bitumen emulsion test?
3. What are the common methods followed for preparation of emulsion?
4. What are the advantages of bitumen emulsion?
5. What are the different types of bitumen emulsion?
6. What are the different tests for bitumen emulsions?
7. How is the cutback bitumen obtained?
8. What are the types of cutback bitumen?
9. What are the tests to be performed on cutback bitumen?
10. What are the properties and requirements for bitumen emulsion?
11. What are the properties and requirements for cutback bitumen?
12. What are the applications of bitumen emulsion?
13. What are the applications for cutback bitumen?

CHAPTER 32

Tests on Polymer and Rubber Modified Bituminous Binder

32.1 OBJECT

To perform tests on polymer and rubber modified bituminous binders.

32.2 THEORY

During hot weather the bituminous pavement surface course becomes soft while during cold weather the bituminous pavement surface course becomes too stiff and brittle. When the weather is hot there is permanent deformation and when the weather is cold there is early cracking. The property of modified bitumen is such that it offers better resistance to deformation when temperature is high and it remains flexible and elastic at low temperatures. Some of the materials used as modifiers of bitumen binders are polymers styrene-butadiene-styrene (SBS), styrene-butadiene rubber (SBR), ethylene vinyl acetate (EVA) and rubber, which are crumb rubber and natural rubber. Binders which are polymer modified are used to improve resistance against deformation, resistance against fatigue and increases in durability in bitumen mixes. Crumb rubber modified bitumen binders improve flexibility, offer resistance to deformation and resistance to deformation and resistance to cracking.

32.3 CLASSIFICATION OF MODIFIED BITUMEN BINDERS

The polymer and rubber modified bitumen are Classified into four types as follows:

1. *Type A: PMB (P):* Plastomeric thermoplastic based
2. *Type B: PMB (E):* Elastomeric thermoplastic based
3. *Type C: N RMB:* Natural rubber and SBR latex based
4. *Type D: CRMB:* Crumb rubber/treated crumb rubber based.

   Types A, B and C are further classified into three grades according to their penetration value.

   The grades of Type A PMB are as follows:

   i. PMB (P) 120 – Type A PMB (P) having a penetration value between 90 and 150
   ii. PMB (P) 70 – Type A PMB (P) having penetration value between 50 and 90.
   iii. PMB (P) 40 – Type A PMB (P) having penetration value between 30 and 50.

   Similarly, the grades of Type B PMB (E) are: PMB (E) 120, PMB (E) 70 and PMB (E) 40. The various grades of types NRMB are: NRMB 120, NRMB 70 and NRMB 40. The grades of type D CRMB are CRMB 50, CRMB 55 and CRMB 60.

In case of CRMB, the numeral indicates the minimum desirable softening point value, viz. CRMB 50 means CRMB having softening point value 5 C minimum.

32.4 TEST ON MODIFIED BITUMINOUS BINDERS

32.4.1 Electric Recovery Test

In this test the dose of polymeric additive in bitumen is optimised and the quality of PMB in laboratory is assessed. The specimen of modified bitumen is elongated upon 10 cm deformation in a ductility machine. The elastic recovery of elongated thread of modified bitumen is compared. It is quantified as elastic recovery. Refer Chapter 29 for details of elastic recovery test.

32.4.2 Separation Test

There is possibility in case of modified bitumen that the modifier used in preparation of the modified bitumen may separated out during storage and transportation. The separation of modifiers during hot storage condition is found by comparison of ring and ball softening point values of the samples. The samples are drawn from top end bottom of conditioned soaked tube of the modified bitumen. The difference in softening points of the respective top and bottom samples is defined as the separation test value. Refer Chapter 28 for details of this test.

32.4.3 Fraass Breaking Point Test

It is the temperature at which bitumen first becomes brittle as seen by the appearance of cracks when a thin film of the bitumen on a metal plaque is cooled. Refer Chapter 27 for details of this test.

32.4.4 Effects of Heat and Air by Thin Film Oven Test

The bituminous binders harden when exposed to atmosphere. The test specimen is subjected to accelerated aging process by 'Thin Film Oven Test'. The amount of hardening of the bituminous material is evaluated from the reduction in penetration test value. It is expressed as a percentage of the original penetration value.

32.4.5 Complex Modulus Test

The resistance to deformation of the binder in the viscoelastic region is defined by complex modulus arid phase angle. The performance aspect of modified bitumen, where elastic recovery is insignificant is evaluated by complex modulus and phase angle. The test method is summarised as follows:

i. Preparation of test specimen.
ii. Placement of the specimen in rheometer maintained at desired test temperature.
iii. Selection of appropriate strain value and operation using the software.
iv. Reporting the recorded values of complex modulus e_s^* and phase angle (sin δ).

32.4.6 Choice of Modified Bituminous Binder

If heavy traffic is on the road and pavement is under overloading conditions, modified bitumen binders are essential. It is the atmospheric temperature at the site of the project on which the selection criteria for the type and grade of modified binder are based. The softest recommended grade is PMB 120 or CRMB 50 which is used for cold climate area. PMB 70 or CRMB 55 is used for moderate climate and PMB 40 or CRMB 60 is used for hot climate areas.

Selection criteria for PMB and CRMB-based on atmospheric temperature as per IRC.

		Maximum atmospheric temperature		
		<30	30 to 40	>40
Minimum atmospheric temperature C	<−15	PMB-120 CRMB-50	PMB-70 CRMB-55	PMB-70 CRMB-60
	15 to −15	PMB-120 CRMB-55	PMB-70 CRMB-55	PMB-40 CRMB-60
	>15	PMB-70 CRMB-55	PMB-40 CRMB-60	PMB-40 CRMB-60

QUESTIONS

1. Classify the different types of bitumen modifiers.
2. Write the advantages of using modified bituminous binders.
3. Write the procedure of elastic recovery test.
4. Write the procedure of Fraass breaking point test.
5. Write the procedure of separation test.
6. Write the advantages of polymer and rubber modified bitumen.
7. List the tests for modified binders.
8. Write the selection criteria for rubber and polymer modified bitumen.
9. Write the application of elastic recovery test.
10. Write the application of Fraass breaking point test.
11. Write the application of separation test.
12. Write the requirements of modified binders.

CHAPTER 33

Water Sensitivity Test on Compacted Bituminous Mixes

33.1 OBJECT

To perform water sensitivity test on compacted bituminous mixes.

33.2 THEORY

The presence of moisture between the aggregates and bitumen causes loss of bond or stripping. The factors on which loss of bond or stripping occur are characteristics of aggregates, binder and anti-stripping additives; properties of compacted bituminous mix; construction procedure, drainage, environment and traffic. There are several laboratory tests which are in use. These tests assess the moisture susceptibility of compacted bituminous paving mixtures. These methods follow the principle such as determination of strength, conditioning another set of identical test specimen and finding the strength ratio.

33.3 APPARATUS

- Vacuum container
- Vacuum pump
- Manometer
- Theoretical maximum specific gravity and density of bitumen paving mixes.
- Freezer maintained at 1.8 C ± 3 C
- Two water baths
- Steel loading strips.

33.4 PROCEDURE

Preparation of test specimens with specified air voids.
1. Prepare a few trial specimen. Vary the amount of compaction with specified air voids content of 7.0 ± 0.5.
2. According to mix design the void content becomes much lower. The initial void content will be in the range of about 6.5 + 07.5%.

33.4.1 Water Sensitivity Test

1. With the help of Marshall compactor, prepare cylindrical bituminous mix specimens. The compared specimen should have air voids of 7 ± 0.5%.
2. At room temperature, allow the compacted specimen to cool. Determine volume and weight of each specimen and then calculate bulk specific gravity and air void content.

3. Prepare two subsets. Each subset should have three test specimen. Prepare specimens such that the two subsets have average void ratio approximately equal.

4. Denote one set of three specimen as 'control specimen' and subject the other set to conditioning.

5. Place control specimen in water bath maintained at 25 ± 0.5 C for two hours. Subject the 'control specimen' under indirect tensile strength test at temperature 25 ± 0.5 C.

6. Keep the other subset of test specimens to conditioning as follows:

   a. Submerse the three specimens in vacuum container filled with water at room temperature. Apply a vacuum of 13–67 kN/m² absolute pressure for 30 minutes. Remove the vacuum and then leave the specimens submersed in water for 5–10 minutes.

   b. Wrap the thin plastic film around each saturated specimen and then place the wrapped specimen in a plastic bag containing 10 ml of water. Seal the plastic bag.

   c. Keep the piastic bag in freezer at a temperature of 18 ± 3 C for a minimum period of 16 hours. Remove the specimens from freezer and place in water bath maintained at 60 ± 1 C for 24 hours.

   d. Remove the specimens from hot water bath and then place under water at room temperature. Remove the plastic bag and plastic film from each specimen and then place the specimens in water bath maintains at 25 ± 0.5 C for two hours.

33.4.2 Indirect Tensile Strength

Determine the indirect tensile strength of all the three control specimens and three conditioned specimens at 25 ± 0.5 C after removing the specimens from water bath, as per the procedure given below:

i. Measure the thickness of each specimen.

ii. Place the specimen in the Marshall testing machine or a compression testing machine and then apply vertical load on the specimen at the rate of 50 mm per minute. Record the maximum compressive strength and continue loading until a vertical crack appears in the specimen. Remove the cracked specimen from machine and then estimate visually the damage caused due to moisture by noting the approximate extent of stripped or bare aggregate on, the fractured faces of the specimen.

33.5 OBSERVATIONS AND CALCULATIONS

The tensile strength of each specimen is calculated as follows:

$$S_t = \frac{2000P}{\pi t d}$$

where S = tensile strength KPa (kN/m²)

           P = Maximum load, N

           t = Mean thickness of specimen, mm

           d = diameter of specimen.

$$\text{Tensile strength ratio (TSR)} = \frac{S_2}{S_1}$$

$$\text{Retained tensile strength, \%} = 100 \frac{S_2}{S_1}$$

where S_1 = Average tensile strength of the control or unconditioned subset of specimens, KPa

S_2 = Average tensile strength of the conditioned subset of specimens, KPa.

33.6 APPLICATIONS

The minimum specified value of retained tensile strength is 80% as per MORTH specification. It is also suggested in the above specification that the water sensitivity test may be conducted if the aggregate fail to satisfy minimum retained coating of art in the 'stripping value test'.

QUESTIONS

1. What are the factors on which loss of bond or stripping depend?
2. What are the equipment for water sensitivity test on compacted bituminous mixes?
3. How would you prepare test specimen with specified air voids?
4. Write in brief water sensitivity test.
5. How are the indirect tensile strength calculated?
6. Write the applications of water sensitivity test.
7. Write expression for tensile strength and write the meaning of different terms.

CHAPTER 34

Benkelman Beam Deflection

34.1 SCOPE

To determine the rebound deflection of a pavement under a standard wheel load and tyre pressure using Benkelman beam.

34.2 THEORY

Benkelman beam is a less expansive deflection measuring equipment. The method of study is simple and easy to carry out. In order to conduct Benkelman beam deflection studies, the Indian Roads Congress has issued guidelines. Benkelman beam measures the magnitude of rebounded deflection of a flexible pavement.

34.3 APPARATUS

Refer IRC 81: 1997

- *Standard Benkelman Beam:* It consists datum frame, rear leg with quick adjustment, clamp, dial gauge, front legs with adjustment screws, hinge of probe beam and probe end of beam.
- *Loaded Truck:* The truck load is adjusted such that the load on rear axle is 8170 kg, equally distributed over the two sets of dual wheels and either side with 4085 kg load on each pair of dual wheels.
- *Tape:* 30 meter length.
- *Chalk:* To mark on pavement surface.
- *Thermometer:* To measure temperature up to 100 C accurate to 1 C.
  Figure 34.1 shows the Benkelman beam.

34.4 PROCEDURE

Refer IRC 81: 1997

1. Take care that the longitudinal spacing between the deflection observation points is not more than 50 m. Consider 20 points in each stretch.
2. Calibrate the Benkelman beam to ensure that the dial gauge and the beam are working correctly.
3. Drive the truck slowly parallel to edge and then stop. Place the left dual wheel tyres of the rear axle centrally over the first measurement of deflection point.
4. Insert the probe end of beam between the gap of the dual wheel and place it exactly over deflection observation (Fig. 34.1).

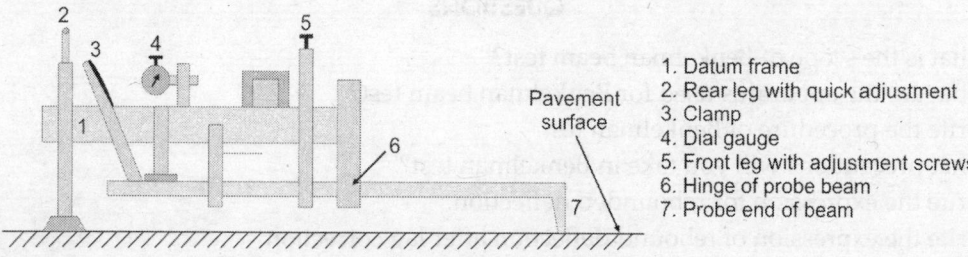

1. Datum frame
2. Rear leg with quick adjustment
3. Clamp
4. Dial gauge
5. Front leg with adjustment screws
6. Hinge of probe beam
7. Probe end of beam

Pavement surface

Fig. 34.1: Benkelman beam

5. Note the initial dial gauge reading D, when there is no change in reading of the dial gauge.
6. Move the truck slowly through a distance of 2.7 m from the deflection observation point and then stop. Note down the intermediate dial gauge reading D_i'.
7. Drive the truck forward through a further distance of 9.0 meter and then note down the final dial gauge reading D_f'.
8. D_o, D_i and D_f are the three deflection dial gauge readings corresponding to one deflection point.
9. Move the truck to next deflection observation point and repeat the test and note the set of three deflections D_o, D_i and D_f.
10. Note down the pavement temperature when the total thickness of bitumen layer is more than 75 mm.

34.5 OBSERVATIONS AND CALCULATIONS

The actual rebound deflection value of pavement surface, D as indicated by probe of the deflection beam is double the value indicated by dial gauge readings D_o, D_f. Therefore the actual rebounded deflection value D is obtained by multiplying by two, the difference between the initial and final dial gauge readings noted from the dial gauge readings, e.g.

$$D = 2(D_o - D_f)$$

Leg Correction

There will be no leg correction (D_i, D_f) if is <0.025 mm. If (D_i, D_f) is >0.025 mm leg correction is applied.

Case (I) If $(D_i, D_f) \leq 0.025$

$$D = 2(D_o, D_f) = 0.02\,(D_o, D_f)\text{ mm}$$

Case (II) If $(D_i, D_f) > 0.025$

$$D = 0.02\,(D_o - D_f) + 0.0582\,(D_i, D_f)\text{ mm}$$

34.6 PRECAUTIONS

1. Protect the beam from temperature variations during test by use of sunshines.
2. Calibrate the beam before the deflection measurement to ensure that the beam and the dial gauge are working properly.
3. Keep spacing between tyre walls equal to 30–40 mm so that beam can be inserted freely.

1. What is the scope of Benkelman beam test?
2. What are the apparatus used for Benkelman beam test?
3. Write the procedure of Benkelman test.
4. What precautions will you take in Benkelman test?
5. Write the expression for rebounded deflection.
6. Write the expression of rebound deflection after lag correction.

CHAPTER 35

Hveem Stabilometer Test

35.1 SCOPE

This method of test covers the procedure for determining the resistance to deformation of compacted bituminous mixture by measuring the lateral pressure developed from applying a vertical load by means of the Hveem stabilometer.

35.2 THEORY

Franics N. Hveem, Materials and Research Engineer for the California Division of Highways advanced and developed the concept for designing paving mixtures based on stabilometer test.

35.3 APPARATUS

- Hveem stabilometer.
- Oven, capable of maintaining a temperature of 60 ± 2 C.
- Initial displacement cylinder.
- Load transfer ram.
- *Measuring device:* Dial indicator assembly for determining the rate of travel of testing machine
- Stopwatch.
- *Testing machine:* 4536 kg capacity minimum.

35.4 PROCEDURE

1. Test specimen at 60 ± 2 C.
2. Place the test specimen in the stabilometer. Make sure that the specimen is firmly seated level on the base.
3. Place load transfer ram on the top of specimen and then adjust pump to give a horizontal pressure of 0.35 kg/cm^2.
4. Start vertical movement of testing machine base at speed of 1.27 mm per minute and record the stabilometer gauge reading when the vertical loads are 227 kg, 454 kg, 907 kg, 1361 kg, 1814 kg, 2268 kg and 2722 kg.
5. Stop the vertical loading exactly at 2722 kg and immediately reduce the load to 454 kg. Turn the displacement pump so that the horizontal pressure is reduced to exactly 0.35 kg/cm^2. Turn the pump handle at approximately two turns per second until stabilometer gauge reads 7 kg/cm^2.

6. Record the number of turns indicated on the dial as the displacement of the specimen. The turns indicator dial reads **0254 mm** and each **254 mm** is equal to one turn.

7. Repeat the procedure for each specimen to be tested.

35.5 OBSERVATIONS AND CALCULATIONS

$$S = \frac{22.2}{[(p_h \times D)/(P_v - P_h)] + 0.0222}$$

where S = Hveem stability value
P_v = Vertical pressure (2758 KPa), i.e. 27.58 kg/cm^2
P_h = Horizontal pressure (P_h is taken at the instant P_v is 2758 KPa, 27.58 kg/cm^2
D = Displacement of specimen.

QUESTIONS

1. Write the scope of Hveem stabilometer test.
2. What are the apparatus used for Hveem stabilometer test?
3. Write the procedure of Hveem stabilometer test.
4. What is the expression for Hveem stability value?
5. Write theory of Hveem stabilometer test.
6. What is the function of measuring device?
7. At what temperature specimen is tested?
8. What are the precautions to be taken for Hveem stabilometer test?
9. What is the rate of vertical movement of testing machine base?
10. At what loads the stabilometer gauge readings are recorded?
11. At what kg the vertical loading is stopped?
12. What is the number of turns per second for pump handle?

CHAPTER 36

Skid Resistance Test

36.1 OBJECT

To determine skid resistance value.

36.2 THEORY

The skid resistance tester, i.e. the pendulum skid resistance tester determines the skid resistance value from actual road surface. The polish stone value and the texture of the aggregate are the two parameter on which skid resistance value depends. The average of a set of reading is taken on the test specimen of aggregate after specified time period of polishing.

36.3 APPARATUS

A friction tester which consists of the following:
- Pendulum arm
- Rubber slider
- Carrying handle
- Swing indicator
- Pendulum arm holder
- Height adjustment screw
- Levelling screws
- Contact gauge.

36.4 PROCEDURE

1. Inspect the road and choose the section to be tested.
2. Set the tester on the road surface in the track chosen so that the slider swings in the direction of traffic.
3. Sweep the road surface ensuring it is free from loose grit.
4. Lower the head of the tester so that the slider just touches the road surface and clamp in position.
5. Lower the pendulum arm until the slider just touches the surface first on one side of vertical and then on the other side.
6. When the arm is vertical, it should not touch the road surface.
7. Spread water over the contact area with hand or a brush.
8. Keep the contact area and slider wet with clean water.

133

9. Release the pendulum and pointer from the horizontal position and note the reading of pointer.

10. Bring the pointer round to its stop. Release the pendulum arm by pressing button. Catch it on the return swing before the slider strikes the road surface.

11. Repeat the test and record the mean of the five successive readings provided they do not differ by the more than three units. If the range is greater than three units, repeat the swings until three successive readings are constant. Record this value.

12. Take mean at each of five locations in the test track spaced at approximately 5–10 m intervals. This mean reading gives a representative value of the skidding resistance of the road.

13. When testing completed at each location, recheck the zero adjustment and sliding length adjustment of the equipment.

14. Measure the temperature of water lying on the road surface immediately after test.

36.5 OBSEVATIONS AND CALCULATIONS

The distance the head (slider) travels after striking the sample is determined by the friction of the surface of the sample (bituminous mixture as surface course) as follows:

Coefficient of friction under specified conditions

$$\mu = \frac{W_x}{P_L}(\cos \delta - \cos \varphi)$$

where $\quad x$ = effective length of the pendulum.

δ = angle of the pendulum to the horizontal end of swing.

φ = angle of the pendulum to the vertical at the beginning of the swing.

W_x = weight of the pendulum

P = total normal load exerted to the specimen

L = length of the specimen traversed by the slider.

36.6 PRECAUTIONS

1. Calibrate the apparatus, i.e. the pendulum skid tester properly.
2. Check the surface of rubber slider for longibility.
3. Wet the pavement surface and rubber slider with water for each test.
4. Note the temperature of pavement.

QUESTIONS

1. What is skid resistance value?
2. Differentiate between skidding and slipping.
3. What is the apparatus used for skid resistance test?
4. Write the components of pendulum skid resistance tester.
5. Write the procedure of skid resistance test.
6. What precautions will you take while conducting skid resistance test?
7. Write the expression for coefficient of friction for skid resistance test.

CHAPTER 37

Dynamic Cone Penetrometer Test

37.1 OBJECT

To perform dynamic cone penetrometer test.

37.2 THEORY

In order to evaluate the properties of the materials at site, such as subgrade soil and the materials below pavement, without the need to cut open the pavement layers, the dynamic cone penetrometer is widely used in the field. In this method a hammer of known weight is allowed to fall on a rod fitted with a metal cone of standard size and shape at the bottom. The resistance to penetration is measured in terms of the depth of penetration due to dropping the hammer of desired weight from a desired height.

37.3 APPARATUS

- Steel rod of diameter 15.8 mm with a replaceable cone tips which has included angle of 60 and a diameter of 20 mm at the base.
- Hammer of weight 8 kg with a height of fall of 575 mm and also a hammer of 4.6 kg weight to be used for weak materials.
- A graduated rod or a graduated vertical scale in increments of 1.0 mm.
- Disposable cone tips and extraction jack if disposable cone trips are not considered.

37.4 PROCEDURE

1. Note the initial reading of the graduated drive rod to the nearest mm.
2. Release the hammer such that it falls from a height of 575 mm and note the penetration reading. Penetration reading gives the value per hammer below.
3. Record the penetration readings and the corresponding number of hammer blows in increments of about 10 mm penetration.
4. Decide the number of blows between each reading depending upon the extent of penetration. Take the reading at every 5 to 10 blows for hard pavement layers like granular sub base/stabilized layers. For weak layers record readings for every blow or two blows.
5. Use the extension rod of dynamic cone penetrometer if the total depth of penetration is >400–500 mm.

37.5 OBSERVATIONS AND CALCULATIONS

Record the data of number of blows and the penetration values. Find the cumulative number of blows and the cumulative penetration values from the data recorded. Plot a graph with cumulative values of number of blows on the x-axis and depth of penetration in mm on the y-axis. The dynamic cone penetrometer value of a layer of material is the penetration value in mm per blow at that depth. The slope of the plot represents the strength characteristic of material in the layer. The change in penetration rate of change in slope of the curve indicates change in material type.

The dynamic cone penetrometer value has been correlated to California Bearing Ratio (CBR) value. So that the results can be used for permanent design.

The equations recommended by some of the organisations are as follows:

Use corps of the engineers:

$$CRB = \frac{292}{(DCPV)^{1.12}} \text{ for soils except for } CL \text{ and } CH \text{ having } CBR \text{ value} <10\%$$

where $DCPV$ is the penetration per blow:

For CL soils with $CBR < 10\%$

$$CBR = \frac{1}{(0.17019 \times DCPV)^2}$$

For CH soils

$$CBR = \frac{1}{(0.00287 \times DCPV)}$$

TRRL of UK (Vide Road Note 8 with 60 cone).

$$\log_{10} CBR = 2.48 - 1.057 \log_{10} DCPV$$

$$(\text{mm/blow})$$

QUESTIONS

1. What is the object for dynamic cone penetration test?
2. What are the equipment for dynamic cone penetrometer test?
3. Write the procedure for dynamic cone penetration test.
4. What is the weight of hammers for dynamic cone penetration test?
5. What is the height of fall of hammer in dynamic cone penetration test?
6. What is the dynamic cone penetration value of a layer of material?
7. What does the slope of the plot between cumulative value of number of blows on x-axis and depth of penetration in mm on y-axis mean?
8. Write the relation between DCP (dynamic cone penetration) value and CBR (California bearing ratio) given by some of the organisations.

CHAPTER 38

Tests on Bituminous Pavement Layer

38.1 OBJECT

To perform tests on bituminous pavement layer.

38.2 THEORY

The dense bituminous macadam, semi-dense bitumen concrete and bituminous concrete. Mixes are designed in laboratory so that it fulfills the desired requirements. These desired requirements are bitumen content, voids in mineral aggregates, density, voids filled with bitumen, air voids stability and flow rule. Large capacity hot mix plants (bituminous mixing plants) prepare the well designed mixes. By the paving machine at temperature of hot mix the bituminous mix is spread and laid to the required thickness grade and camber. The specified set of rollers at temperature range of the mix compacts this layer. Care should be taken to minimise segregation of the bituminous mix at each and every stage. Required compaction is provided so as to achieve the mix properties close to those of the designed mix.

38.3 CORE DRILLING METHOD FOR EXTRACTING TEST SAMPLES

Initial Preparation

Mark the locations of coring and number of the locations of coring. If second acceptance test core becomes necessary, locate it at a distance of 100 mm from the edge of the first core. In order to reduce the possibility of damage to core specimen, cool the layer with dry ice for 10 to 15 minutes before coring.

Backfill the core hole soon after the core samples are extracted from the pavement.

38.3.1 Equipment

1. *Core drilling machine:* A portable power drilling machine with water cooling system.
2. *Core barrels:* Core barrels with thin wall with diamond cutting edges having diameter 100 and 150 mm and length up to 250 mm or more it required.
3. *Core retrieval tool:* It extricates cores from pavement.
4. *Other equipment:* Arrangements for packing, preservation and transportation of core samples.

38.3.2 Test Procedure

1. Position the core drill at desired location such that the core barrel is perpendicular to pavement surface.

2. Turn on the machine along with cooling unit. Lower steadily the core barrel by rotating.

3. Turn off the barrel rotation unit and cooling unit when the core drill cuts pavement up to required depth. Take out core barrel and remove the core sample. Wipe dry core sample immediately. Mark the core sample its identification number. Keep the sample in an insulated cool chamber.

38.4 LABORATORY TESTS ON CORE SAMPLES

38.4.1 Density Test

Carry out density test first on reference core samples.

Additional samples are collected and tested, collect the sample at a distance of 100 mm from the edge of the first core. Determine the volume of each core sample from the mean diameter and height and then find density by knowing the weight of the sample.

38.4.2 Void Analysis

For void analysis refer the Marshall stability test and mix design.

38.4.3 Bitumen Extraction and Bitumen Content Introduction

The bitumen extraction test is conducted by collecting sample from core drilling after conducting the density voids tests.

38.4.4 Apparatus

- Motorised centrifuge
- Extraction apparatus
- Fitter rings
- Suitable solvent (e.g. commercial grade benzene digital balance of capacity 500–1000 g).

38.4.5 Procedure

1. Take about 500 g of representative sample of bitumen mix. Dry the sample well and then take its weight.

2. Place the sample in extraction apparatus covered with benzene and allow it to disintegrate in the solvent for about 1 hour.

   Dry the filter ring with filter, weigh and fit on the rim of the bowl. Place the container under the extraction apparatus to collect the extract.

3. Start the centrifuge machine and let it revolve at slow speed and then speed is increased to a maximum of 3600 rpm. Maintain this speed till no further solvent flows down the drain.

4. Stop the machine and then add 200 ml solvent in the extraction apparatus and then operate the centrifuge machine as before.

5. Repeat the process of adding 200 ml solvent and extraction of the bender there or more number of times till the extract is clear and lighter in colour than that of the straw.

6. Remove the filter ring, dry in air and then in a thermostatically control led oven at 115 C to constant weight and then weigh accurately.

38.5 OBSERVATIONS AND CALCULATIONS

The percentage bitumen binder content in terms of the weight of total bitumen mix is calculated as follows:

$$\text{Binder contents \%} = \frac{(W_1 + W_4 W_2 W_3)}{W_1} \times 100$$

Where
W_1 = Weight of sample of bituminous mix
W_4 = Increase in weight of the filter ring
W_2 = Weight of sample after extraction
W_3 = Weight of fine material recovered from the extract

QUESTIONS

1. What are the different pavement mixes?
2. What are the desired requirements for tests on bituminous pavement layer?
3. Write initial preparation for core drilling method for extracting test samples.
4. Write equipment used for core drilling method for extracting test samples.
5. Write the procedure of core drilling method for extracting test samples.
6. What are the laboratory tests on core samples?
7. What are the equipment for laboratory tests on core samples?
8. Write the procedure for laboratory tests one core samples.
9. Write the expression for percentage binder content and write the meaning of terms used.

CHAPTER 39

Unevenness Measurements by Bump Integrator and MERLIN

39.1 OBJECT

To measure unevenness by bump integrator and MERLIN.

39.2 THEORY

The functional requirements of orad users are fulfilled by construction and maintenance of road pavements. Only few users are aware of the increase in vehicle operation cost due to pavement undulations. There is increase in fuel consumption when the vehicles are subjected to vertical oscillations when the longitudinal profile of pavement surface is uneven. The undulation or unevenness of pavement surface are of three types as shown below:

i. Rough surface profile with minor corrugation.

ii. Unevenness surface with large number of undulations.

iii. Surface having large size depressions on some stretches.

The evaluation of undulations and unevenness in pavements can be done by following methods:

i. Methods which are based on physical measurement of the surface undulations.

ii. Methods which are based on in direct measurements.

iii. Methods which are based on subjective assessment.

39.3 MEASUREMENT OF UNEVENNESS INDEX BY BUMP INTEGRATOR

39.3.1 Equipment

The bump integrator is a trailer unit which consists of a single automobile wheel with rubber tyre of specified size mounted on a heavy chassis through suitable bearings and a suspension system which is hauled at a uniform speed of 30 kmph by a towing vehicle. The bump integrator is a response type road roughness measuring equipment.

39.3.2 Procedure

1. Identify the stretch of the road to be tested and mark the start and end points by bold lines across the pavement width.

2. Make the unevenness measurements along the normal wheel path, by making the test runs such that the test wheel of the bump indicator runs along the desired wheel path.

3. Make the first test run along the left wheel path on the onward trip from the starting point to the end, while conducting the tests on undivided roads. In the return trip, the test wheel is run along the other wheel path.
4. Set to zero the digital unity/counters, one indicating the cumulative value of undulations and other indicating the number of revolutions of the test wheel. Note down when the test wheel crosses the starting line or else note the initial reading.
5. Note and record the readings of both the counters when the test wheel of bump indicator unit crosses the ending line.
6. Reverse the hauling vehicle and the bump integrator and make the test run along the other wheel path in the counter run noting down the initial and final readings.
7. Similarly make total three to four test runs along each wheel path so that mean value of unevenness is determined.

39.3.3 Observations and Calculations

The unevenness index or roughness index for the test run or stretch of any length is obtained as follows:

$$\text{Unevenness Index (UI) (mm/km)} = \frac{10XY}{W}$$

where　X = Bump integrator readings from field after setting initial reading to zero.
　　　　Y = Number of revolutions of the test wheel per km.
　　　　W = Number of wheel revolutions from the field.

The maximum permissible values of unevenness index or roughness index measured with a bump integrator for different surfaces as per the Indian Road Congress specifications are as follows:

Sl. No.	Condition of road surface		
	Good	Average	Poor
1. Surface dressing	<3500	3500–4500	>4500
2. Open graded pre-mix carpet	<3000	3000–4000	>4000
3. Mix seal surfacing	<3000	3000–4000	>4000
4. Semi-deuse bitumen concrete	<2500	2500–3500	>3500
5. Bituminous concrete	<2000	2000–3000	>3000

39.4 MEASUREMENT OF UNEVENNESS BY MERLIN

MERLIN is a simple equipment which can measure the unevenness of pavement surface more accurately, but at a slow speed. MERLIN denotes the short form of 'Machine for Evaluating Roughness using low-cost instrumentation. MERLIN has two feet which are spaced at 1.8 m and a probe that rest on the wheel track. The probe lies midway between two feet.

Procedure

1. Mark the wheel path along which the readings are to be taken. Keep the MERLIN at starting point by moving the MERLIN.

2. Record the location of the pointer on the chart with a cross at the appropriate column and keep a record of total number of observations.
3. Raise the handle of MERLIN, so that only the wheel is in contact with the road surface. Move it to forward to the next measuring point and repeat this process.
4. Locate the next point after each revolution of the wheel of MERLIN. Point a mark on the rim of the wheel and then take measurement everytime. Rotate the wheel such that the mark comes to the road surface.
5. Take at least 200 readings at regular intervals or for 200 wheel revolutions.
6. Remove the chart from MERLIN when 200 observations are made. Count the number of cross marks from either end.
7. Measure the spacing between two marks D in millimeter and take as the roughness on the MERLIN scale.

39.5 OBSERVATIONS AND CALCULATIONS

The relation between the MERLIN scale and international roughness index by bump integrator are given as follows:

International roughness index (IRI)   = 0.593 + 0.0471D

where D is the roughness in terms of MERLIN scale and is measured in mm.

QUESTIONS

1. What fulfills the functional requirements of the road user?
2. How cost is affected due to pavement undulations?
3. What causes increase in fuel consumption in vehicles?
4. What are the categories of undulations or unevenness of pavement surface?
5. What are the three broad classifications of evaluation of undulations or unevenness in the pavements?
6. What are the equipment for measurement of unevenness index by bump integrator?
7. Write the procedure for measurement of unevenness index by bump integrator.
8. What is a bump integrator?
9. Write expression for unevenness index.
10. What are the maximum permissible values of unevenness index or roughness index measured with a bump integrator for different surfaces as per the Indian Road Congress specifications?
11. What are the equipment for measurements of unevenness by MERLIN?
12. What is meaning of MERLIN?
13. What is MERLIN?
14. Write the procedure for measurements using MERLIN.
15. Write expression for international roughness index.
16. What is relationship between MERLIN scale and international roughness index by bump integrator?

Appendices

Important Value of Aggregates (Crushing Value, Impact Value Abrasion Value and Flakiness Index)

Specified limits of percent aggregate crushing value for different types of road construction

Type of road construction	Aggregate crushing value not more than
1. Flexible pavements	
i. Soling	50
ii. Water-bound-macadam	40
iii. Bituminous macadam	40
iv. Bituminous surface dressing or thin per mix carpet	30
v. Dense-mix carpet	30
2. Rigid Pavements	
i. Other than wearing course	30
ii. Surface or wearing course	45

Classification based on aggregate impact value

Aggregate impact value	Classification
< 10%	Exceptionally strong
10–20%	Strong
10–30%	Satisfactory for road surfacing
>35%	Weak for road surfacing

Maximum aggregate impact value as per Indian Road Congress

S. No.	Type of pavements	Maximum aggregate impact value %
1.	Bituminous surface dressing penetration macadam, bituminous carpet concrete and cement concrete wearing course.	30
2.	Bitumen-bound-macadam, base course	35
3.	WBM base course with bitumen surfacing	40
4.	Cement concrete base course	45

Aggregate abrasion value according to Indian Road Congress

S. No.	Type of pavement layer	Maximum permissible abrasion value in %
1.	Water bound macadam, sub base course	60
2.	WBM base course with bituminous surfacing	50
3.	Bituminous bound macadam	50
4.	WBM surfacing course	40
5.	Bituminous penetration macadam	40
6.	Bituminous surface dressing, cement concrete surface course	35
7.	concrete surface course	30

Typical Los Angeles abrasion value for common rock types

Rock type	Los Angles abrasion value
Basalt	10–17
Diabase	13–21
Dulurnite	1 8–30
Gneiss	33–57
Gnenite	27–49
Limestone	19–30

Aggregate crushing and impact value of various rocks

Rock group	Aggregate crushing value	Aggregate impact value
Basalt	14	15
Granite	20	19
Limestone	24	23
Quartzite	16	21

Maximum allowable flakiness index of aggregates in different types of pavement construction

S. No.	Type of Pavement construction	Maximum limits of flakiness index, %
1.	Bitumen carpet	30
2.	i. Bituminous/Asphaltic concrete	
	ii. Bituminous penetration macadam 1 m	
	iii. Bituminous surface dressing (single coat, two coats and per coated)	
	iv. Built-up spray grout	25
3.	i. Bituminous macadam	
	ii. Water bound macadam, base and surfacing courses	25

Properties of Bitumen and Tar

Range of softening point as per Indian Standard Institution for various grades of bitumen

Bitumen grades	Softening point, C
*A25 and 35	55–70
*S35	50–65
A45, S45 and A65	45–60
S65	40–55
A90 and S90	35–50
A200 & S200	30–45

* A Denotes bitumen from Assam Petroleum
* S Bitumen from sources other than from Assam Petroleum

Specifications for test temperature and range of viscosity for road tar (according to IS: 215–1981)

Road tar type	RT-1	RT-2	RT-3	RT-4	RT-5
Orifice size, mm	10	10	10	10	10
Test temperature	35 C	45 C	35 C	55 C	65 C
Viscosity is sec.	30–55	30–55	35–60	35–70	35–70

Specifications for test temperature and range of viscosity for cutback bitumen (according to IS: 217–1961)

Grades—SC, MC and RC	0	1	2	3	4	5
Orifice size, mm		4	4	10	10	10
Test temperature C	25	25	25	25	40	40
Viscosity in sec.	25–75	50–150	10–20	26–27	14–45	60–140

Minimum ductility for various grades of bitumen (according to ISI)

Source of paving bitumen and penetration grade		Minimum ductility value cms
Assam Petroleum	A25	5
	A35	10
	A45	12
A65, A90 and A200		15
Bitumen from sources other than Assam Petroleum S35		50
S45, S65 and S90		75

Marshall's Mix Properties, Skid Resistance Value and Polished Value

Desired criteria for Marshall mix properties

Property	Bitumen binder	Aggregate	Air voids
Stability	Just enough to coat the aggregate	Angular shapes rough texture dense graded	
Durability	Need a lot of bitumen to coat the aggregate and completely fill out the voids	Capable to resist weathering effect	Need minimal air void to prevent oxidation and entrance of water
Flexibility skid resistance	Just enough to coat the aggregate	Open graded aggregate containing many minerals in it	Need some void space to prevent bleeding

Suggested minimum value of skid resistance

Category	Type of site	Minimum value
A	Difficult sites such as i. Round abouts ii. Bends with radius less than 150 m iii. Gradients 1 in 20 or steeper of lengths, >100 m iv. Approaches to traffic lights	65
B	Heavy trafficked roads (carrying 75,000 vehicles per day)	55
C	All other sites	45

Polished stone value of some rock groups

Rock group	Average	Range
Artificial (slag)	0.59	0.35–0.74
Basalt	0.62	0.45–0.81
Hornfels	0.45	0.40–0.45
Porphyry	0.56	0.43–0.71
Limestone	0.43	0.30–0.75
Granite	0.59	0.45–0.70
Quartzite	0.58	0.45–0.67
Grit stone	0.72	0.60–0.82

Requirements of Bitumen Emulsion and Cutback Bitumen

Requirements for bitumen emulsion (cationic type)

S. No.	Characteristics	Grade of emulsion					
		SS-1	SS-2	MS	SS-1	SS-2	I.S. method of test
i.	Residue on 600 micron IS sieve (percent by mass, max)	0.05	0.05	0.05	0.05	0.05	8887-B
ii.	Viscosity by saybolt furol viscometer, seconds						3117
	At 25 °C	–	–	–	20–100	30–150	
	At 50°C	20–100	100–300	50–300	–	–	
iii.	Coagulation of emulsion at low tempera (applicable to locations where the ambient temperature is below 15°C)	Nil	Nil	Nil	Nil	Nil	8887-C
iv.	Storage stability after 24 hours percentage, max	2			2	2	8887-D
v.	Particle charge	Positive	Positive	Positive	Weak Positive	Positive	8887-E
vi.	Coating ability and water resistance						
	Coating, dry aggregate	–	–	Good	–	–	
	Coating, after spraying	–	–	Fair	–	–	
	Coating, wet aggregate	–	–	Fair	–	–	
	Coating after spraying	–	–	Fair	–	–	
vii.	Stability to mixing with cement (percent coagulation), max	–	–	–	2	2	8887-G
viii.	Miscibility with water	No coagulation	No coagulation	No coagulation	–	No coagulation	

Contd...

Contd...

S. No.	Characteristics	Grade of emulsion					
		SS-1	SS-2	MS	SS-1	SS-2	I.S. method of test
ix.	Tests on residue						
	(a) Residue by evaporation	60	67	65	50	60	8887–1
	(b) Penetration at 25°C/100 g/5 sec	80–150	80–150	60–150	60–350	60–120	1203
	(c) Ductility, 27°C/cm, min	50	50	50	50	50	1208
	(d) Solubility: in trichroethylene, min	98	98	98	98	98	1216
x.	Distillation in percent, by volume at:						
	(a) 190°C	–	–	–	20–55	–	
	(b) 225°C	–	–	–	30–75	–	
	(c) 260°C	–	–	–	40–90	–	
	(d) 315°C	–	–	–	60–100	–	
xi.	Water content, percent by muss, max	–	–	–	20	–	

Requirements of rapid curing (RC) cutback bitumen

S. No.	Characteristics	RC 70		RC 250		RC 800		RC 3000		IS method of test
1	2	3		4		5		6		7
i	Kinematic viscosity at 60°C, cSt	70	140	250	500	800	1600	3000	6000	IS:1206 (Part 3)-1978
ii	Flash point, pensky martens closed type °C	26	–	26	–	26	–	26	–	IS: 1209–1978
iii	Distillate volume percent of total distillate up to 360°C									
	a. Up to 190°C	10	–	–	–	–	–	–	–	
	b. Up to 225°C	50	–	35	–	15	–	–	–	
	c. Up to 260°C	70	–	60	–	45	–	25	–	
	d. Up to 315°C	85	–	80	–	75	–	70	–	
iv	Residue from distillation up to 360°C, percent by volume (by difference)	55	–	65	–	75	–	80	–	
v	Test on residue from distillation up to 360°C									
	a. Viscosity at 60°C, poises	600	2400	600	2400	600	2400	600	2400	IS:1206 (Part 3) 1978
	b. Ductility at 27°C, cm	100	–	100	–	100	–	100	–	IS:1208–1978
	c. Matter soluble in trichloroethylene, percent bymass	99	–	99	–	99	–	99		IS: 1216–1978
vi	Water content, percent by weight	–	0.2	–	0.2	–	0.2	–	0.2	IS: 1211–1978

Requirements of medium curing (MC) cutback bitumen

S. No.	Characteristics	MC 30		MC 70		MC 250		MC 800		MC 3000		Method of test
1	2	3		4		5		6		7		8
i.	Kinematic viscosity at 60°C St	30	60	70	140	250	500	800	1600	3000	6000	IS:1206 (Part 3)1978
ii.	Flash point, pensky martens closed type °C	38	–	38	–	65	–	65	–	65	–	IS:1209–1978
iii.	Distillate volume percent of total distillate up to 360°C											
	a. Up to 225°C	–	25	–	20	–	10	–	–	–	–	
	b. Up to 260°C	40	70	20	60	15	55	–	35	–	15	
	c. Up to 315°C	75	93	65	90	60	87	45	80	15	75	
iv.	Residue from distillation up to 360°C, percent by volume (by difference)	50	–	55	–	67	–	75	–	80	–	
v.	Test on residue from distillation up to 360°C											
	a. Viscosity at 60°C, poises	30	1200	300	1200	300	1200	300	1200	300	1200	IS:1206 (Part 3) 1978
	b. Ductility at 27°C, cm	100	–	100	–	100	–	100	–	100	–	IS 1208–1978
	c. Matter soluble in trichloroethylene, percent bymass	99	–	99	–	99	–	99	–	99	–	IS 1216–1978
vi.	Water content, percent by weight	–	0.2	–	0.2	–	0.2	–	0.2	–	0.2	IS 1211–1978

Requirements of medium curing (SC) cutback bitumen

S. No.	Characteristics	SC 70	SC 250	SC 800	SC 3000		Method of test
1	2	3	4	5	6		7
i.	Kinematic viscosity, cSt	70 140	250 500	800 1600	3000 6000		IS:1206 (Part 3)–1978
ii.	Flash point, pensky martens closed type °C	65 –	79 –	93 –	107 –		IS: 1209–1978
iii.	Total distillate up to 360°C volume percent	10 30	4 20	2 12	– 5		IS: 1203–1978
iv.	Kinematic viscosity on distillation residue up to 360°C	4 70	8 100	20 160	40 350		
v.	Test on residue for distillation up to 360°C						
	(a) Residue of 100 percent	50 –	60 –	70 –	80 –		IS 1204–1978
	(b) Ductility of 100 penetration residue at 27°C cm	100 –	100 –	100 –	100 –		IS 1208–1978
	(c) Matter soluble in trichloroethylene, percent by mass	99 –	99 –	99 –	99 –		IS 1216–1978
iv.	Water content, percent by mass	– 0.2	– 0.2	– 0.2	– 0.2		IS 1211–1978

Bibliography

1. Arora KK. *Soil Mechanics and Foundation Engineering,* Standard Publishers & Distributors, Delhi, 2003.
2. Bishop AW, Henkel DJ. *The Measurement of Soil Properties in the Triaxial Test,* Eduward Amolde, 1962.
3. Casagrande A. *Classification and Identification of Soils,* Trans, ASCE, Vol. 113, 1948.
4. Das BM. *Advanced Soil mechanics,* McGraw- Hill Book Co., New York, 1985.
5. Gulati SK. *Engineering Properties of Soils,* Tata McGraw-Hill Publishing Co. Ltd., New Delhi, 1978.
6. IRC 37: *Guidelines for the Design of Flexible Pavement,* The Indian Road Congress, New Delhi, 2001.
7. IRC 81: (1st Revision), *Guidelines for Strengthening of Flexible Road Pavement using Benkalman Beam Deflection Technique,* The Indian Road Congress, New Delhi, 1997.
8. IS 73: *Indian Standard Specification for Paving Bitumen,* The Bureau of lndian Standard, New Delhi, 1992.
9. IS 73: *Indian Standard Specification for Paving Bitumen,* The Bureau of Indian Standards, New Delhi, 2006.
10. IS 1202: *Indian Standard Specification for Determination of Specific Gravity,* The Bureau of lndian standards, New Delhi, 1978.
11. IS 1203: *Determination of Penetration,* 1978.
12. IS 1205: *Indian Standard Methods for Testing Tar and Bituminous Materials. Determination of Softening Point,* The Bureau of Indian Standards, New Delhi, 1978.
13. IS 1206 (Part III): 1978, *Indian Standard Specification for Determination of Viscosity,* The Bureau of Indian Standards, New Delhi.
14. IS 1206: *Indian Standard Methods for Testing Tar and Bituminous Materials: Determination of Ductility,* The Bureau of Indian Standards, New Delhi, 1978.
15. IS 1209: *Indian Standard Methods for Testing Tar and Bituminous Materials: Determination of Flash and Fire Point of Bitumen,* The Bureau of Indian Standards, New Delhi, 1978.
16. IS 1212: *Indian Standard Methods for Testing Tar and Bituminous Materials: Determination of Loss on Heating of Bitumen,* The Bureau of Indian Standards, New Delhi, 1978.
17. IS 1216: (1st Revision), *Indian Standard Specification for Determination of Solubility in Carbon Disulphide or Trichloroethylenes,* the Bureau of Indian Standards, New Delhi, 1978.
18. IS 1498: *Classification and Identification of Soils for General Engineering Purposes,* The Bureau of Indian Standards, New Delhi, 1970.
19. IS 2384 (Part V): *Method of Test for Aggregate for Concrete: Soundness,* The Bureau of Indian Standards, New Delhi, 1963.
20. IS 2720 (Part 4): *Method of Test for Soils, Part 4: Grain Size Analysis,* 1985.
21. IS 2720 (Part 5): *Method of Test for Soils. (Part 5) Determination of Liquid and Plastic Limit* (Second Revision), Reaffirmed 1990, 1985.
22. IS 2720 (Part 6): *Method of Test for Soils, (Part 6) Determination of Shrinkage Factors* (First Revision) (Amendment 1), Reaffirmed 1990, 1972.
23. IS 2720 (Part 7): *Method of Test for Soils: (Part 7) Determination of Water Content Dry Density Relation using Light Compaction* (Second Revision) (Amendment 2), Reaffirmed 1987, 1980.

24. IS 2720 (Part 8): *Method of Test for Soils: (Part 8) Determination of Water Content Dry Density Relation using Heavy Compaction.* (Second Revision) Reaffirmed 1990, 1983.

25. IS 2720 (Part 16): (2nd Revision) *Laboratory Determination of CBR,* The Bureau of Indian Standards, New Delhi, 1987.

26. IS 2720 (Part 28): *Method of Test for Soils: (Part 28) Determination of Dry Density of Soils in Place by Sand Replacement Method* (First Revision) Reaffirmed 1988, 1974.

27. IS 2720 (Part 29): *Method of Test for Soils: (Part 29) Determination of Dry Density of Soil in Place by Core-cutter Method* (First Revision) Reaffirmed 1988, 1975.

28. IS 6241: *Method of Test for Determination of Stripping Value of Road Aggregate,* 1971.

29. IS 9214: *Method of Determination of Modulus of Sub Grade Reaction of Soils in Field,* The Bureau of Indian Standards, 1974.

30. IS 9381: *Indian Standard Methods for Testing Tar and Bituminous Material: Determination of Fraass Breaking Point of Bitumen,* The Bureau of Indian Standards, New Delhi, 1978.

31. Khanna, Justo. *Highway Material Testing Laboratory Manual,* Nern Chand and Bross Roorker (UP), 1977.

32. Lambe TW. *Soil Testing for Engineers,* Wiley Eastern, New Delhi, 1977.

33. Murthy VNS. *Textbook of Soil Mechanics and Foundation Engineering,* CBS Publisher and Distributors, New Delhi, 2007.

34. Punmia BC, Jain AK and Jain AK. *Soil Mechanics and Foundations,* Lakshmi Publications (P) Ltd. New Delhi, 2005.

35. Ranjan G, Rao ASR. *Basic and Applied Soil Mechanics,* New Age International (P) Ltd., New Delhi, 2000.

36. Rao GV. *Principles of Transportation and Highway Engineering,* Tata McGraw-Hill, New Delhi, 2000.

37. Road Research Laboratory, *Bituminous Material in Road Construction,* Her Majesty's Stationary Office London, 1963.

38. Salt GF, Szatkowski WS. *A Guide to Levels of Skidding Resistance for Road TRRL,* Report No. LR510, Transportation and Road Research Laboratory, Department of Transportation, Crow thom, UK 1973.

39. Singh A. *Soil Engineering in Theory and Practice,* Asian Publishing House, Bombay.

40. Terzaghi K. *Theoretical Soil Mechanics,* Wiley, 1943.

41. Terzaghi. *Theory to Practice in Soil Mechanics,* Wiley and Sons, New York, 1960.

42. Terzaghi K, Peck RB. *Soil Mechanics in Engineering Practice,* John Wiley and Sons.

43. Traxler Ralph N. *Asphalt Its Composition, Properties and Uses,* Chapman Hall Ltd., London, 1961.

44. Wallace and Martin. *Asphalt Pavement Engineering,* McGraw-Hill, 1975.

45. Wu TH. *Soil Mechanics,* Allyn and Boston, Inc., 1996.

46. Yong H Hung, *Pavement Analysis and Design,* Prentice Hall, 1993.

47. Yoder EJ. *Principles of Pavement Design,* John Wiley, 1975.

48. Yong RN, Win Terkorn BP. *Soil Properties and Behaviour,* Elsevier Scientific Publishing Co. Amsterdam, 1975.

Index

155